HEAT CONVERSION SYSTEMS

Georg Alefeld
Technical University of Munich
Munich, Germany

and

Reinhard Radermacher
University of Maryland
College Park, Maryland, U.S.A.

CRC Press
Taylor & Francis Group
Boca Raton London New York

CRC Press is an imprint of the
Taylor & Francis Group, an **informa** business

CRC Press
Taylor & Francis Group
6000 Broken Sound Parkway NW, Suite 300
Boca Raton, FL 33487-2742

ISBN 13: 978-0-367-44975-9 (pbk)
ISBN 13: 978-0-8493-8928-3 (hbk)

Visit the Taylor & Francis Web site at
http://www.taylorandfrancis.com

and the CRC Press Web site at
http://www.crcpress.com

Library of Congress Card Number 93-829

Library of Congress Cataloging-in-Publication Data

Alefeld, Georg.
 Heat conversion systems / by Georg Alefeld and Reinhard
Radermacher.
  p. cm.
 Includes bibliographical references and index.
 ISBN 0-8493-8928-3
 1. Thermodynamics. 2. Heat engines. I. Radermacher, Reinhard.
II. Title.
TJ265.A43 1993
621.402—dc20

93-829
CIP

Introduction

This text discusses a wide range of heat conversion systems. Emphasis is placed on those systems that are based on Rankine cycles. The Rankine cycle is characterized by phase transitions like evaporation/condensation and, as shown later, by desorption/absorption. Included in this group are absorption and vapor compression heat pumps and refrigeration machines, heat transformers (reversed absorption heat pumps), sorption heat storage devices (i.e. thermo-chemical energy storage), and chemical heat pumps. These systems represent networks in which working fluids and possibly absorbent fluids or other auxiliary fluids circulate while undergoing certain thermodynamic processes. The networks are composed of certain base components such as compressors or expansion machines, pumps, expansion valves, condensers, evaporators, absorbers, desorbers, resorbers, and heat exchangers for internal heat recovery. With the design of any such heat conversion system, an optimization problem has to be solved. Whatever the design objective might be, usually a combination of the following requirements has to be fulfilled:

(A) Highest efficiency for a given investment

(B) Minimum first cost for a desired efficiency

(C) Best possible match of internal parameters (such as the concentration, pressure, temperature of the chosen working fluids) to external requirements (such as temperatures of useful heat, waste heat, or refrigeration load)

(D) User friendly start up and shut down characteristics and control concepts

(E) High reliability, low maintenance

Design objectives take into account the relative weight of these requirements. Addition or deletion of a single connection between base components may significantly alter properties of a circuit and may result in a new configuration. Up-to-date cycles are optimized by experience based on state-of-the-art configurations. It is necessary to compare *all* suitable configurations to be certain that the best possible configuration for a given design objective is found. Therefore, the task of finding the best cycle for any application is preceded by the more difficult and more fundamental task of discovering all meaningful configurations and combinations of base components of heat conversion systems. This is a problem in combinatorics under consideration of the First and Second Laws.

This text will discuss three procedures for finding the optimal heat conversion system for a particular application. The first procedure starts from main components like heat and mass exchangers and compressors. The second procedure providing the fastest and most revealing results starts from basic cycles and will be outlined in detail. A third procedure offers an interesting overview using group theory.

Tailored for graduate students in the field of energy conversion, refrigeration, heating and air conditioning, applied physics, and mechanical and chemical engineering, this text originated from the combined material of the second part of a course, "Energy-Conversion, Storage and Transport", taught at the Technical University in Munich, Germany (printed lecture notes "Waermeumwandlungssysteme", G. Alefeld, 1983) and from the course, "Advanced Energy Conversion Systems", taught at the University of Maryland in College Park, Maryland, U.S.A. While the prerequisite for understanding this material is basic thermodynamics as found in existing literature, the contents of this text are basically new (except for occasional reviews) or can only be found in specialized publications. Since this text is based on lecture notes, redundancy is not completely eliminated.

By applying the First and Second Laws of Thermodynamics, a classification scheme of the possible operating modes of heat transformation systems is developed in Chapter One. The hurried reader may focus only on Section 1.4 of this chapter. In Chapters Three and Four, the thermodynamic processes of simple absorption and vapor-compression heat pumps are discussed.

Chapter Five, which is of special importance, introduces rules for the design of more complicated energy conversion systems. The rules are formulated so that the composition of new cycles avoids unnecessary entropy production. The information and design rules represent the basis for Chapters Six through Ten. When using the rules, many new configurations of heat conversion systems can be discovered. Special representations of cycle configurations are created, producing simple graphs which inherently include the First and Second Laws of Thermodynamics and the conservation of mass. These representations lead to a very simple design method of new heat conversion cycles from which the machinery required to operate the cycles can be deduced. The contents of Chapter Five can best be understood from examples given when the rules are applied. After skimming through Chapter Five, the reader is encouraged to go to Section 1 of Chapters Seven and Nine and then back to Chapter Five. The most important results of this text are documented in the specialized graphs of Figures 7.1 (Chapter Seven) and 9.1 (Chapter Nine).

Chapters Eight and Ten, dealing with configurations of three and more stages, may be skipped initially. Chapter Seven, Section 11 and Chapter Nine, Section 12, which introduce the application of permutation operators to the organization of basic components, are also of aesthetic value. Chapter Eleven discusses solid/vapor systems. These can be used as single or multi-stage heat pumps and heat transformers. As any sorption system, they also have the capability of storing exergy which can be used for the production of refrigeration, heat, or work. This storage capability is the thermal equivalent to electro-chemical storage.

Chapter Twelve deals with open sorption and storage systems.

Large sections of this text can be understood as a research and development program for years to come. Essential progress in heat conversion processes can be expected in two areas:

(A) New configurations of components and thermodynamic cycles which become feasible as improved fluids or components become available

(B) Improved working fluids and working fluid combinations

These areas and their interconnection are discussed in detail in this text. Research in energy conversion systems is in its infancy and the field is open to those who wish to pursue further technical and scientific research in this area.

The authors are hopeful that this text conveys new knowledge of heat conversion systems to students and interested readers and that this knowledge will serve as an invitation to the reader to pursue further technical and scientific research. The field of energy conversion systems is of economic and ecologic importance and only at the beginning of being explored.

The authors are very grateful to Felix Ziegler and Günther Feuerecker for providing the problems and solutions. The problems are chosen from several courses to provide for the students' benefit the best understanding of the material. The completion of this text was supported by the U.S. Department of Energy, which is gratefully acknowledged.

Table of Contents

Chapter One
THE THEORY OF HEAT TRANSFORMATION

1.1 Introduction 1
1.2 Two Temperature Levels 2
1.3 Three Temperature Levels 6
1.4 Many Temperature Levels 12

Chapter Two
WORKING FLUIDS

2.1 Working Fluids in Thermodynamic Diagrams 19
2.2 Properties of Working Fluids 29
2.3 Research Areas for New Fluids 36

Chapter Three
VAPOR-COMPRESSION HEAT PUMPS AND POWER PLANTS

3.1 The Rankine Cycle with Input and Output of Work 37
3.2 Vapor-Compression Heat Pumps, Refrigeration Units, and Heat Engines 38
3.3 Condensation and Evaporation of Working Fluid Mixtures 46
3.4 Rankine Sorption Cycles with Input and Output of Work 50
3.5 Vapor-Compression Heat Pump or Heat Engine with Solution Circuit 52
3.6 Transformation of Availability by Pairs of Exchange Units 62
3.7 COP of Compressor Heat Pumps Based on the Second Law 62

Chapter Four
ABSORPTION HEAT PUMPS (REFRIGERATORS)
AND HEAT TRANSFORMERS

4.1 Principle of Operation 83
4.2 The Absorption Heat Pump, Combination of Heat Engine and Heat Pump 83
4.3 Description of the Complete Absorption Cycle 85
4.4 Coefficients of Performance 90
4.5 Means for Increasing the COP 98
4.6 Limitations of Single-Stage Absorption Heat Pumps 111
4.7 Comparison of the Performance of Various Systems 111
4.8 COP Evaluation With Limited Data 113

Chapter Five
RULES FOR THE DESIGN OF MULTISTAGE, ABSORPTION HEAT PUMPS,
HEAT TRANSFORMERS, VAPOR-COMPRESSION HEAT PUMPS,
HEAT ENGINES, AND CASCADES

5.1 Introduction 123
5.2 Design Rules for the Configuration of Heat Conversion Systems 126
5.3 Graphic Representations 128

Chapter Six

CASCADE CONFIGURATIONS 133

Chapter Seven

TWO-STAGE ABSORPTION HEAT PUMPS AND HEAT TRANSFORMERS

7.1   Classification of Two-Stage Absorption Units 135
7.2   Operating Modes of Two-Stage Absorption Units with Six Exchange
        Units and Two Working Fluid Streams 137
7.3   Internal Heat Exchange 154
7.4   High Efficiencies (Heat Ratios) 158
7.5   Large Temperature Lifts 161
7.6   Heat Transformer 163
7.7   Heat Pump Transformer 165
7.8   Heating and Cooling Capabilities 165
7.9   Connections for Working Fluid and Solution Streams 168
7.10   Switching between Configurations of Different Classes 170
7.11   Set Properties 174

Chapter Eight

THREE AND MULTISTAGE ABSORPTION CONFIGURATIONS

8.1   Operating Modes of Three-Stage Units with Five Temperature Levels 179
8.2   Three-Stage Absorption Configurations 181
8.3   Multistage Absorption Configurations 186
8.4   Evaluation of the COP of Multistage Absorption Cycles 187

Chapter Nine

TWO-STAGE CONFIGURATIONS WITH
COMPRESSION/EXPANSION MACHINES

9.1   Classification of Two-Stage Absorption-Compression/Expansion Machines 207
9.2   Special Modes of Operation 208
9.3   Adjustment of Coefficients of Performance and Large Temperature Lifts 218
9.4   Adjustment of Pressure Levels 219
9.5   Adjustment of Temperature Lift 220
9.6   Simultaneous Compressor and Absorption Operation 220
9.7   Special Operating Ranges for Four and More Temperature Levels 221
9.8   Internal Heat Exchange 222
9.9   Refrigeration, Heating, and Power Generation 225
9.10   Using Superheat 225
9.11   Switching Between Configurations of the Same Class 227
9.12   Set Properties 228

Chapter Ten

MULTISTAGE ABSORPTION/COMPRESSION/EXPANSION MACHINES

10.1   Two-Stage Compression/Expansion Machines 231
10.2   Multistage Combinations 231

10.3 Three-Stage Configurations with One and Two
Compression/Expansion Machines 233
10.4 Special Multistage Configurations 236

Chapter Eleven
SOLID-VAPOR HEAT PUMPS, REFRIGERATORS, HEAT TRANSFORMERS, AND HEAT STORAGE DEVICES

11.1 The Operating Principle of Sorption Storage Devices 241
11.2 Sorption Unit for Storage of Availability 243
11.3 The Zeolite Heat or Cold Storage Device 244
11.4 Multistage Storage Devices, Storage Heat Pumps, and Heat Transformers 246
11.5 Quasi-Continuous Operation with Solid-Vapor Systems 250
11.6 Solid-Vapor Systems with Compression/Expansion Machines 251

Chapter Twelve
OPEN HEAT TRANSFORMATION AND STORAGE SYSTEMS

12.1 Open Heat Conversion System for Steam Generation 255
12.2 Open Storage Systems 255

PROBLEMS 257

APPENDIX ONE 275

APPENDIX TWO 283

INDEX 289

CHAPTER ONE

The Theory of Heat Transformation

A heat conversion system is any machinery that accepts heat from heat reservoirs on certain temperature levels and rejects heat to heat reservoirs on other temperature levels. The production or consumption of work may be part of the heat conversion process. One example is the heat engine. It operates on only two temperature levels. High temperature heat is supplied, low temperature heat rejected, and work produced.

The processes involved in the conversion of heat are governed by the First and Second Laws of Thermodynamics and can be irreversible or can approach reversibility to a certain degree.

This chapter presents a synopsis of operating modes of heat conversion systems. A classification system of the operating modes is developed.

At first, simple heat conversion systems with two temperature levels for the heat reservoirs are discussed. By extending to more temperature levels the general rules about the numbers of reversible and irreversible modes of operation for heat conversion systems can be derived.

Although this chapter is quite basic, it can be skipped by the hurried reader and consulted when needed.

1.1 INTRODUCTION

An example of a simple heat conversion system may be a heat engine. This heat engine can be generalized to include more than two temperature levels and heat reservoirs to which heat is rejected by the engine or from which heat is supplied to the engine. Depending on the application, a heat reservoir of a certain temperature level T_i can act as a heat source or a heat sink. Whether heat is supplied or rejected at temperature level i is indicated by the sign of the heat Q_i. The following sign convention is used. Heat and work are considered positive ($Q_i > 0$, $W > 0$) when they are supplied to the heat conversion system. Figure 1.1 pictures a heat conversion system capable of exchanging heat with a given number of heat reservoirs of fixed temperatures. The system may produce work (heat engine), require the input of work (heat pump), or require the input of heat with only parasitic power (absorption heat pump). Depending on the number of heat reservoirs and the sign of the work, various combinations of signs for the heat flows Q_i at the temperature levels T_i are possible. Certain combinations of signs are in violation of the First or Second Law of Thermodynamics. Other combinations represent modes of operation which are inherently irreversible. The combinations of particular interest for technical applications are those which can, at least in principle be operated reversibly. These modes of operations (solutions) are now derived by applying the First and Second Laws of Thermodynamics. The solutions will be the basis for the classification scheme for heat conversion systems. The First and Second Laws may be written as follows:

$$\text{First Law} \qquad \sum_{i=0}^{n} Q_i + W = 0 \qquad\qquad (1.1)$$

$$\text{Second Law} \qquad \sum_{i=0}^{n} \frac{Q_i}{T_i} \leq 0 \quad \begin{matrix} \text{irr} \\ \text{rev} \end{matrix} \qquad\qquad (1.2)$$

Q_i and W may be understood as quantities of heat and work or their time derivatives in stationary systems. The First Law represents the energy balance for the entire heat conversion system. The symbol $\leq$ in the equation for the Second Law results because the heat supplied

1

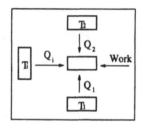

Figure 1.1 Exchange of heat and work.

to the conversion system is considered positive. This is not a contradiction to the more common form of the Second Law which states for the reservoirs:

$$\Delta S \geq 0 \qquad (1.3)$$

The increase in entropy ΔS can be calculated by:

$$\Delta S = -\sum \frac{Q_i}{T_i} \qquad (1.4)$$

Again, the signs are based on the convention shown above. The numbering sequence for the heat Q_i is chosen so that for the respective temperatures T_i the following relationship holds:

$$T_0 < T_1 < ...T_i < T_n \qquad (1.5)$$

First, the possible combinations of signs for two and three heat reservoirs are studied. Based on these results, the later extension to many heat reservoirs on different temperature levels will become clear. To indicate in the following figures whether or not a certain combination of signs is allowed according to the First and Second Laws, the following symbols are used:

I   Sign combination excluded by First Law,

II Sign combination excluded by Second Law,

i   Sign combination permitted, the cycle operates only irreversibly,

r   Sign combination permitted, the cycle can, in principle, operate reversibly, but also irreversibly.

1.2 TWO TEMPERATURE LEVELS

The matrix in Figure 1.2 represents all permutations of the signs for the two heat flow Q_0 and Q_1.

1.2.1 Solutions for Processes Without Work (W = 0, $T_0 < T_1$)

Referring to Figure 1.2, the following cases can be distinguished:

Cases 1 and 3: These cases are not allowed by the First Law. The heat conversion system would either reject heat on all temperature levels or absorb heat on all temperature levels.

Case 2: This sign combination fulfills the First and Second Laws. The Second Law is fulfilled only with the symbol < allowing only an irreversible process. This is demonstrated by the following equation:

	Case			
	1	2	3	4
Q_1	+	+	-	-
Q_0	+	-	-	+

Figure 1.2 Sign matrix for two temperature levels.

	Case			
	1	2	3	4
Q_1	+	+	-	-
Q_0	+	-	-	+
	I	i	I	II

Figure 1.3 Sign matrix for two temperature levels. The last line indicates whether the solution is excluded by the First Law (I), the Second Law (II), or possible only irreversibly.

$$\left.\begin{array}{l} Q_1 - Q_0 = 0 \\[2mm] \dfrac{Q_1}{T_1} - \dfrac{Q_0}{T_0} \le 0 \end{array}\right\} \rightarrow \dfrac{1}{T_1} - \dfrac{1}{T_0} \le 0, \text{ because } T_0 \le T_1 \tag{1.6}$$

Heat is supplied at T_1 and rejected at T_0, without the production of work. Case 2 represents a conventional heat exchanger. The entropy generation amounts to:

$$\Delta S = -\sum \frac{Q_i}{T_i} = Q_i \frac{T_1 - T_0}{T_1 T_0} \tag{1.7}$$

Case 4: This sign combination fulfills the First Law, but violates the Second Law which can be derived as follows:

$$\left.\begin{array}{l} -Q_1 + Q_0 = 0 \\[2mm] -\dfrac{Q_1}{T_1} + \dfrac{Q_0}{T_0} \le 0 \end{array}\right\} \rightarrow -\dfrac{1}{T_1} + \dfrac{1}{T_0} \le 0, \rightarrow T_1 \le T_0 \tag{1.0}$$

The combination of signs for the two temperatures T_0 and T_1 that results from the application of the Second Law contradicts the assumption $T_0 < T_1$.

In the last row of the sign matrix of Figure 1.3, the symbols that were defined at the end of the last chapter are displayed.

1.2.2 Solutions for $W \ne 0$

The First and Second Law for $W \ne 0$ read:

$$Q_0 + Q_1 + W = 0, \tag{1.9}$$

$$\frac{Q_0}{T_0} + \frac{Q_1}{T_1} \le 0 \tag{1.10}$$

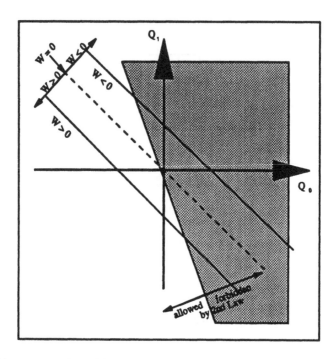

Figure 1.4 Graphic representation of the relationship between work and heat as defined by the Laws of Thermodynamics.

Figure 1.4 shows a graphic representation of the relationship between work and heat for the First and Second Laws for two temperature levels. The x-axis represents Q_0 and the y-axis Q_1. The four quadrants represent the four cases of sign combinations of Figure 1.2. With no work input or output, the solution allowed by the First Law is $Q_0 = Q_1$. This solution is represented by the dashed line. Below and above the dashed line are the ranges for $W > 0$ and $W < 0$, respectively. The dashed line has a slope of 1.0 and so have the lines marked "1st Law" for $W < O$ and $W > O$. The Second Law is fulfilled for all points on (reversible) and to the left (irreversible) of the line marked "2nd Law". The slope of the Second Law line is negative and depends on the ratio of the temperatures T_1 and T_0. To the right of that line is the range that is excluded by the Second Law, shaded area in Figure 1.4. As can be seen from Figure 1.5, there are four permitted cases. Case 4 and Case 6 can, in principle, be reversible. Cases 3 and 2 are always irreversible. Figure 1.5 shows the sign combinations for two temperature levels with work input and output.

From Equations 1.9 and 1.10, the following relationships can be derived after elimination of either Q_0 or Q_1.

$$\Delta S = Q_1 \frac{T_1 - T_0}{T_1 T_0} + \frac{W}{T_0} \tag{1.11}$$

or

$$\Delta S = -Q_0 \frac{T_1 - T_0}{T_1 T_0} + \frac{W}{T_1} \tag{1.12}$$

Inserting these relationships into Equation 1.11 confirms that, for the reversible case, the entropy production is zero.

1.2.3 Solutions for W > 0 (Vapor-Compression Heat Pump)

The condition $W > 0$ limits the consideration to the area below the dashed line in Figure 1.4. Case 4 is discussed first, since it includes the reversible vapor-compression heat pump.

	Case							
	1	2	3	4	5	6	7	8
W	+	+	+	+	-	-	-	-
Q_1	+	+	-	-	+	+	-	-
Q_0	+	-	-	+	+	-	-	+

Figure 1.5   Sign matrix for two temperature levels and for input and output of work.

Case 4:   Work is supplied to the heat conversion system, high temperature heat is rejected to the reservoir at T_1, and low temperature heat is supplied by the low temperature reservoir. Heat is pumped from T_0 to T_1. The point located on the line marked "2nd Law" represents reversible heat pumps. As the heat pump becomes more irreversible, the cooling capacity Q_0 decreases for constant work input and reaches zero at the limit to Case 3. At the boundary between Cases 3 and 4, $Q_0 = 0$, all work is converted into high temperature heat. The heat pumping action has ceased. The coefficient of performance for heating and cooling can be expressed in the following way:

$$\frac{|Q_1|}{W} \leq \frac{T_1}{T_1 - T_0} \qquad \frac{Q_0}{W} \leq \frac{T_0}{T_1 - T_0} \tag{1.13}$$

Heat Pumping   Refrigeration

The entropy production in this operating range amounts to:

$$0 \leq \Delta S < \frac{W}{T_1} \tag{1.14}$$

Case 3:   Since $Q_0 < 0$ and $W > 0$, the First and Second Laws are fulfilled. The work input to the system is rejected as heat at both temperatures, T_1 and T_0. The boundary between Case 3 and Case 2 is determined by $Q_1 = 0$. The entire work is rejected as heat at T_0. The entropy production in Case 3 amounts to:

$$\frac{W}{T_1} < \Delta S < \frac{W}{T_0} \tag{1.15}$$

It exceeds the entropy production of Case 4.

Case 2:   The sign combination of Figure 1.5 for Case 2 indicates that heat Q_1 in addition to work is supplied. The entire energy is then rejected as Q_0. This solution is permitted by the First Law and the Second Law, but it is highly irreversible.

$$\Delta S > \frac{W}{T_0} \tag{1.16}$$

Case 1:   Case 1 is excluded by the First Law.

1.2.4 Solution for W < 0 (Heat Engine)

In contrast to heat pumps (W > O), for heat engines (W < O) only one sign combination is compatible with the First and Second Laws:

Case 6:   Heat is supplied at T_1 and rejected at T_0 and work is produced. Less reversible heat engines at constant work output require more heat input. The following equation is an expression for the efficiency of the system:

$$\frac{|W|}{Q_1} \le \frac{T_1 - T_0}{T_1} \tag{1.17}$$

The entropy production according to Equation (1.11) is given by:

$$\Delta S = Q_1 \frac{T_1 - T_0}{T_1 T_0} - \frac{|W|}{T_0} \tag{1.18}$$

Case 7:   The sign combination for Case 7 of Figure 1.5 is forbidden by the First Law. The heat conversion system would supply heat at all temperature levels and produce work.

Case 5:   This sign combination is valid according to the First Law but excluded by the Second Law. A system cannot produce work when only heat is supplied without any waste heat rejection.

Case 8:   This sign combination is allowed by the First Law but excluded by the Second Law, for the same reasons as in Case 5. A cycle cannot produce work and high temperature heat when only low temperature heat is supplied.

1.2.5 SUMMARY

For two temperature levels, three sign combinations are permitted for W > 0; one is reversible and irreversible (representing the vapor-compression heat pump), and two are only irreversible. One valid sign combination exits for W < 0, reversible and irreversible, representing the heat engine. For W = 0, there is just one sign combination permitted representing a heat exchanger. An example for how Cases 4, 3, and 2 can technically occur is shown in Figure 1.6. The heat pump operates between T_0 and T_1, but parasitic heat flow occurs between T_1 to T_0

1.3 THREE TEMPERATURE LEVELS

For three temperature levels, there exist $2^3 = 8$ sign combinations for the heat flows, as displayed in Figure 1.7. To find those combinations compatible with the First and Second Laws, the equations for the First and Second Laws are combined by eliminating one of the three Q_i:

$$-Q_2 \left(\frac{T_2 - T_0}{T_2} \right) - Q_1 \left(\frac{T_1 - T_0}{T_1} \right) \le W$$

$$-Q_2 \left(\frac{T_2 - T_1}{T_2} \right) + Q_0 \left(\frac{T_1 - T_0}{T_0} \right) \le W$$

$$+Q_1 \left(\frac{T_2 - T_1}{T_1} \right) + Q_0 \left(\frac{T_2 - T_0}{T_0} \right) \le W \tag{1.19}$$

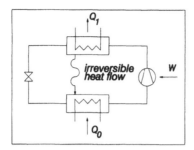

Figure 1.6 Heat pump with irreversible heat flow from condenser to evaporator.

	Case							
	1	2	3	4	5	6	7	8
Q_2	+	+	+	+	-	-	-	-
Q_1	+	+	-	-	+	+	-	-
Q_0	+	-	-	+	+	-	-	+
	I	i	i	r	II	r	I	II

Figure 1.7 Sign matrix for three temperature levels and eight sign vectors.

In these equations all Q_i can be still positive or negative. The coefficients are written such that they are always positive. Note that only three coefficients are Carnot factors.

1.3.1 Solution for W = 0 (Absorption Heat Pump, Heat Transformer)

Equations 1.19 are now applied, and the amounts of heat are substituted with the signs according to the various cases of the matrix in Figure 1.7. The Q_i are now to be understood as absolute quantities, $|Q_i|$. All three equations have to be fulfilled. In the subsequent discussion, that equation which is the most restrictive will be selected.

Case 1:   Excluded by First Law.

Case 2:

$$-Q_2\left(\frac{T_2 - T_0}{T_2}\right) - Q_1\left(\frac{T_1 - T_0}{T_1}\right) \leq 0 \tag{1.20}$$

Since both terms on the left side are negative, this sign combination represents an irreversible solution. Heat is supplied at T_2 and T_1 and rejected at T_0.

Case 3:

$$-Q_1\left(\frac{T_2 - T_1}{T_1}\right) - Q_0\left(\frac{T_2 - T_0}{T_0}\right) \leq 0 \tag{1.21}$$

Again this is a valid combination, but only irreversibly. Heat added at T_2 flows to T_1 and T_0 and is rejected.

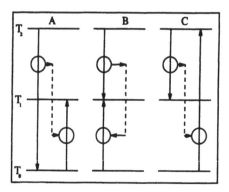

Figure 1.8   Heat transformation over three temperature levels without external work.

Case 5:

$$Q_2\left(\frac{T_2 - T_0}{T_2}\right) + Q_1\left(\frac{T_1 - T_0}{T_1}\right) \le 0 \tag{1.22}$$

Not compatible with Second Law.

Case 7:   Excluded by First Law.

Case 8:

$$Q_1\left(\frac{T_2 - T_0}{T_1}\right) + Q_0\left(\frac{T_2 - T_0}{T_0}\right) \le 0 \tag{1.23}$$

Excluded by Second Law.

Cases 4 and 6:   For both sign combinations, the above equations always show a positive term combined with a negative term, so that reversible as well as irreversible solutions are possible.

Case 4:   Heat is supplied at T_2. Due to its availability, heat is pumped from T_0 to a higher temperature level. The efficiency (heat ratio) can be expressed as shown in Equation 1.24 based on the above Equations 1.19 with $W = 0$.

$$\frac{Q_1}{Q_2} \le \frac{\dfrac{1}{T_0} - \dfrac{1}{T_2}}{\dfrac{1}{T_0} - \dfrac{1}{T_1}} = \frac{\dfrac{T_2 - T_0}{T_2}}{\dfrac{T_1 - T_0}{T_1}} \tag{1.24}$$

Two additional heat ratios can be calculated. Each represents a further heat conversion system, Figure 1.8. Equation 1.24 corresponds to A, 1.25 to B, and 1.26 to C.

$$\frac{Q_1}{Q_2} \le \frac{T_1}{T_2} + \frac{T_2 - T_1}{T_2}\frac{T_1}{T_1 - T_0} \tag{1.25}$$

$$\frac{Q_1}{Q_2} \leq \frac{\dfrac{T_1}{T_2}}{1 - \dfrac{\dfrac{T_2 - T_1}{T_2}}{\dfrac{T_2 - T_0}{T_2}}} \tag{1.26}$$

In Figure 1.8, each circle represents a heat engine or heat pump. When the vertical arrow through the circle points downward, heat flows from the higher temperature level to the lower temperature level, and work is produced. The horizontal arrow symbolizes the work output. When the vertical arrow points upward, heat is pumped from the lower to the higher temperature level and work has to be supplied, indicated by the horizontal arrow pointing to the circle. The cycles shown in Figure 1.8 can be imagined as being contained within the "black box" of Figure 1.1. Figure 1.1 would have a total of three temperature levels. The work generated is completely consumed within the machine itself. There is no exchange of work with the surroundings.

It can be expected that in case of a technical realization, cycle B will exhibit the least irreversibility.

If the process according to Case 4 is used for cooling, the efficiency can be written as:

$$\frac{Q_0}{Q_2} \leq \frac{\dfrac{T_2 - T_1}{T_2}}{\dfrac{T_1 - T_0}{T_0}} \tag{1.27}$$

The denominator is not a Carnot factor. Equation 1.27 has again three versions according to the three cycles in Figure 1.8.

Case 6:   Heat is added at T_1. By employing only a part of its availability with respect to T_0, the temperature of the remaining part of the heat is lifted from T_1 to T_2. This cycle, termed "heat transformer" is just the reverse of the heat pump cycle according to Case 4. The usable heat rejected at T_2 per unit of heat supplied at T_1 amounts to:

$$\frac{Q_2}{Q_1} \leq \frac{\dfrac{1}{T_0} - \dfrac{1}{T_1}}{\dfrac{1}{T_0} - \dfrac{1}{T_2}} \cdot \frac{\dfrac{T_1 - T_0}{T_1}}{\dfrac{T_2 - T_0}{T_2}} \tag{1.28}$$

The symbol "$\leq$" is reversed compared to the equation for the heat pump, Equation (1.24).

This equation has three versions according to the three cycles in Figure 1.8, but the directions of all arrows are inverted. The last line in the sign matrix in Figure 1.7 summarizes the results for $W = 0$: Two sign combinations are valid, reversible, and irreversible, (up to a certain maximum value of irreversibility); two sign combinations are permitted only irreversibly; two are excluded due to the Second Law; two more violate the First Law. With increasing irreversibility, Case 4 turns into Case 3, (for $Q_0 = 0$), and Case 6 into Case 2 (for $Q_2 = 0$). The entropy production for Case 4 is limited as described in Equation 1.29:

$$0 < \Delta S < -\left(\frac{Q_2}{T_2} - \frac{Q_1}{T_1}\right) = Q_2 \frac{T_2 - T_1}{T_1 T_2} \tag{1.29}$$

		Case														
	1	2	3	4	5	6	7	8	9	10	11	12	13	14	15	16
W	+	+	+	+	+	+	+	+	−	−	−	−	−	−	−	−
Q_2	+	+	+	+	−	−	−	−	+	+	+	+	−	−	−	−
Q_1	+	+	−	−	−	−	+	+	+	+	−	−	−	−	+	+
Q_0	+	−	−	+	+	−	−	+	+	−	−	+	+	−	−	+
	I	i	i	r	r	i	r	r	II	r	r	r	II	I	r	II

Figure 1.9   Sign matrix for three temperature levels with the input or output of work.

and the entropy production for Case 6 by:

$$0 < \Delta S < -\left(\frac{Q_1}{T_1} - \frac{Q_0}{T_0}\right) = Q_1 \frac{T_1 - T_0}{T_1 T_0} \tag{1.30}$$

1.3.2 Solutions for W > 0

Figure 1.9 shows the sign matrix for three temperature levels with work input and output. The last row indicates the compatibility with the First and Second Laws.

Case 1:   Excluded by First Law.

Case 2:

$$-Q_2\left(\frac{T_2 - T_0}{T_2}\right) - Q_1\left(\frac{T_1 - T_0}{T_1}\right) \leq W \tag{1.31}$$

This case is permitted, but only irreversibly.

Case 3:

$$-Q_1\left(\frac{T_2 - T_1}{T_1}\right) - Q_0\left(\frac{T_2 - T_0}{T_{11}}\right) \leq W \tag{1.32}$$

This case is permitted, but only irreversibly.

Case 6:

$$-Q_1\left(\frac{T_2 - T_1}{T_1}\right) - Q_0\left(\frac{T_2 - T_0}{T_0}\right) \leq W \tag{1.33}$$

This case is permitted, but only irreversibly.

Cases 4, 5, 7, 8: These sign combinations represent modes of operation which, in principle, can be reversible.

Case 4:   From the inequalities (1.19), the following two inequalities can be derived:

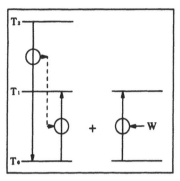

Figure 1.10   Heat transformation (including external work) over three temperature levels.

$$Q_1 \le W \frac{T_1}{T_2 - T_0} + Q_2 \frac{\dfrac{T_2 - T_0}{T_2}}{\dfrac{T_1 - T_0}{T_1}}$$

$$Q_0 \le W \frac{T_0}{T_1 - T_0} + Q_2 \frac{\dfrac{T_2 - T_1}{T_2}}{\dfrac{T_1 - T_0}{T_0}} \tag{1.34}$$

The Equations 1.34 can be interpreted as follows (Figure 1.10): A vapor-compression heat pump (refrigeration unit) operates between T_0 and T_1. In addition, the heat Q_2 is used for heating at T_1 or cooling at T_0. This combination is of interest for compression heat pumps which are driven by an internal combustion engine. The waste heat of the engine fires an absorption heat pump. The evaporator and condenser of the absorption heat pump operate under the same temperature levels as those of the vapor-compression heat pump. With increasing irreversibility, Case 4 turns into Case 3 when $Q_0 = 0$.

Case 5:   From the inequalities (1.19), the following two inequalities can be derived:

$$Q_1 \le W \frac{T_1}{T_1 - T_0} - Q_2 \frac{\dfrac{T_2 - T_0}{T_2}}{\dfrac{T_1 - T_0}{T_1}}$$

$$Q_0 \le W \frac{T_0}{T_1 - T_0} - Q_2 \frac{\dfrac{T_2 - T_1}{T_2}}{\dfrac{T_1 - T_0}{T_0}} \tag{1.35}$$

The equations can be interpreted as follows (Figure 1.11): A vapor-compression heat pump operates between T_0 and T_1. A portion of the heat delivered at T_1 is used to drive a heat transformer which lifts some of the heat to an even higher temperature level T_2. With increasing irreversibility, Case 5 turns into Case 6 for $Q_0 = 0$.

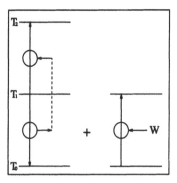

Figure 1.11   Heat transformation with external work over three temperature levels.

Cases 7 and 8 can be interpreted in a similar way, as was done for Cases 4 and 5. With increasing irreversibility, Case 7 turns into 2 when $Q_2 = 0$, while Case 8 turns into 6 for $Q_1 = Q_0 = 0$, or into Case 2 when $Q_2 = Q_0 = 0$.

1.3.3   Solutions for W < 0

If work is produced by a heat conversion system, then there are only four solutions permitted: Cases 10, 11, 12, and 15 can be, in principle, reversible. All other sign combinations are excluded by either the First or the Second Law. It should be noted that there exists a symmetry for the cases of $W > 0$ and $W < 0$. While there exist three irreversible but permitted cases for $W > 0$, there are three cases forbidden by the Second Law for $W < 0$.

1.4   MANY TEMPERATURE LEVELS

By comparing sign vectors (column of the matrix in Figure 1.9, for example) for n, n + 1, n − 1 temperature levels, certain relationships between permitted and forbidden vectors can be recognized. The relationships and the rules for finding the valid solutions are outlined in the following:

1.4.1   Reversible Solutions (W ≠ 0)

The reversible sign combinations for the following set of equations

$$\sum_{i=0}^{n} Q_i + W = 0$$

$$\sum_{i=0}^{n} \frac{Q_i}{T_i} = 0 \qquad T_0 < T_1 < T_2 ... < T_n \tag{1.36}$$

can be found by first developing the solutions of the same set of equations for W=0 but with one additional temperature level T_w. The temperature T_w is assumed to be *higher* than any temperature T_i.

$$\sum_{i=0}^{n} Q_i + W = 0$$

$$\sum_{i=0}^{n} \frac{Q_i}{T_i} + \frac{W}{T_w} = 0 \qquad T_0 < T_1 < T_2 ... < T_n < T_w \tag{1.37}$$

For T_w approaching infinity, the Equations 1.37 turn into Equation 1.36.

Consequently, the reversible solutions for n temperature levels and W ≠ 0 are identical to those for n + 1 temperature levels and for W = 0, if the sign of the heat on the highest temperature level is the same as for the work W.

1.4.2 Reversible and Irreversible Solutions

When W = 0 the following equations hold:

$$\sum_{i=0}^{n} Q_i = 0$$

$$\sum_{i=0}^{n} \frac{Q_i}{T_i} + \Delta S = 0 \qquad \Delta S \geq 0 \qquad T_0 < T_1 < T_2 ... < T_n \tag{1.38}$$

In this case, the sign combinations of reversible and irreversible solutions can be found by developing the reversible solutions for the same set of equations with one additional temperature level:

$$\sum_{i=0}^{n} Q_i + Q_s = 0$$

$$\sum_{i=0}^{n} \frac{Q_i}{T_i} + \frac{Q_s}{T_s} = 0 \qquad T_s < T_0 < T_1 < T_2 ... < T_n \tag{1.39}$$

The temperature T_s is assumed to be lower than any other temperature T_i. When:

$$Q_s = \Delta S T_s \qquad \text{and} \qquad T_s \rightarrow 0, \text{ with}$$

$$\frac{Q_s}{T_s} = \text{const} = \Delta S > 0, \quad \text{i.e.,} \quad Q_s > 0 \tag{1.40}$$

then the set of Equations 1.39 turns into Equations 1.38.

The sign combinations for the reversible and irreversible solutions for n temperature levels are identical to the reversible solutions for n + 1 temperature levels, if the lowest temperature is neglected and only those are chosen with the positive sign on the lowest temperature level. (The latter is a consequence of the condition that $Q_s \geq 0$ because $\Delta S \geq 0$.)

Now the case W ≠ 0 is included in this consideration, and both methods are combined.

$$\sum_{i=0}^{n} Q_i + Q_s + W = 0$$

$$\sum_{i=0}^{n} \frac{Q_i}{T_i} + \frac{Q_s}{T_s} + \frac{W}{T_w} = 0 \qquad T_s < T_0 < T_1 < T_2 ... < T_n < T_w \tag{1.41}$$

When the reversible solutions for n + 2 temperature levels are known, all reversible and irreversible solutions for n temperature levels and W ≠ 0 can be derived.

1.4.3 General Relationships and Tables

In the sign matrixes on the following pages, Figures 1.12 through 1.15, the very left column is valid for W = 0 and n temperature levels. The very right column is valid for W = 0 and n - 1

n=2, W=0 or n=1, W=0				
	1	2	3	4	
Q_1	+	+	−	−	W
Q_0	+	−	−	+	Q_0
	I	i	I	II	
1 Solution				
0 reversible Solutions				

Figure 1.12 Sign matrix for two temperature levels without external work, or one temperature level with external work.

n=3, W=0 or n=2, W=0							
	1	2	3	4	5	6	7	8	
Q_2	+	+	+	+	−	−	−	−	W
Q_1	+	+	−	−	−	−	+	+	Q_1
Q_0	+	−	−	+	+	−	−	+	Q_0
	I	i	i	r	II	i	r	II	
4 Solutions							
2 reversible Solutions							

Figure 1.13 Sign matrix for three temperature levels without external work, or two temperature levels with external work.

Case																
	1	2	3	4	5	6	7	8	9	10	11	12	13	14	15	16	
Q_3	+	+	+	+	+	+	+	+	−	−	−	−	−	−	−	−	W
Q_2	+	+	+	+	−	−	−	−	−	−	−	−	+	+	+	+	Q_2
Q_1	+	+	−	−	−	−	+	+	+	+	−	−	−	−	+	+	Q_1
Q_0	+	−	−	+	+	−	−	+	+	−	−	+	+	−	−	+	Q_0
	I	i	i	r	r	i	r	r	II	r	I	II	r	r	r	II	

Figure 1.14 Sign matrix for four temperature levels without external work, or three temperature levels with external work.

temperature levels. The symbols in the last row of the matrix have the same meaning as described above.

The relationships between the solutions with and without the input or output of work, including irreversibilities, can be verified immediately with the help of Figures 1.12 through 1.15. With increasing entropy production, solutions that are, in principle, reversible, change into solutions that were classified as irreversible solutions.

Note: (a) Solutions with W = 0 which are entirely irreversible convert into, in principle, reversible ones when work is produced.

      (b) Solutions with W = 0 which are excluded by the Second Law convert into, in principle, reversible ones if work is supplied.

By extension from n to n + 1 temperature levels, $T_{n+1} > T_n$, all reversible solutions for n temperature levels remain reversible independent of the sign of Q_{n+1}. The permitted but

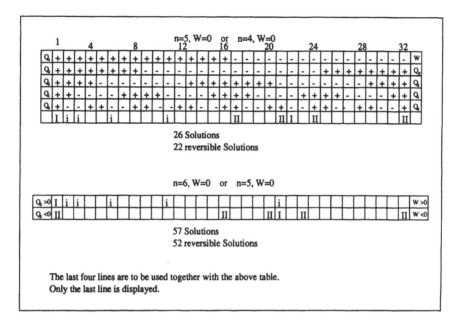

Figure 1.15 Sign matrix for five temperature levels without external work, or four temperature levels with external work.

irreversible combinations for n temperature levels are added to the reversible ones when $Q_{n+1} < 0$. Those solutions that are excluded by the Second Law for n temperature levels become reversible when $Q_{n+1} > 0$.

The following relationships between reversible solutions and the number of temperature levels hold:

No Net Input or Output of Work

Number of sign permutations: 2^n

The number of solutions allowed by the First and Second Law is:

$$2^n \qquad -2 \qquad\qquad -(n-1) \qquad = 2^n - n - 1$$

Excluded by: First Law,   Second Law

The first term in the above equation is the total number of sign vectors. The second term accounts for the fact that only two sign vectors violate the First Law independent of the number of temperature levels. These are the sign vectors which consist of "+" or "−" only. The third term in the above equation accounts for the sign vectors that have all "−" on temperature levels higher than any "+". These vectors violate the Second Law. Their number is $n - 1$. The number of irreversible solutions is the same as the number of solutions that violate the Second Law. These solutions are represented by sign vectors where any "+" is on a higher temperature level than any "−". In these cases, heat flows from a higher temperature to a lower temperature, irreversibly. With this, the number of reversible solutions, Z_{rev} amounts to:

$$Z_{rev}(W = 0) = 2^n - 2 \quad -2(n-1) = 2^n - 2n$$

Net Input of Work

The number of possible sign permutations is $\frac{1}{2} 2^{n+1} = 2^n$.

The number of the solutions allowed by the First and Second Laws:

$$2^n \qquad -1 \qquad\qquad -0 \qquad = 2^n - 1)$$

Excluded by: First Law,   Second Law

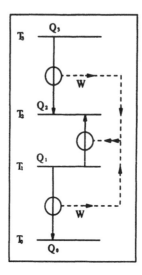

Figure 1.16   Schematic representation of a heat pump transformer.

The first term is the number of possible sign vectors. The second term in the above equation accounts for the fact that one sign vector violates the First Law. The number of sign vectors violating the Second Law is zero. The number of the reversible solutions Z_{rev} amounts to:

$$Z_{rev}(W > 0) = 2^n - 1 - n = 2^n - (n+1)$$

Net Output of Work
The number of sign permutations is $\frac{1}{2} 2^{n+1} = 2^n$.
The number of solutions that fulfill the First and Second Laws is

$$2^n \quad\quad -1 \quad\quad\quad\quad\quad\quad -n \quad\quad\quad = 2^n - (n+1)$$
$$\text{Excluded by: First Law,} \quad\quad \text{Second Law}$$

The number of solutions that previously (for $W = 0$) violated the Second Law is increased by one, that one which violated the First Law. Because all previously irreversible solutions are now reversible, the number of the reversible solutions amounts to:

$$Z_{rev}(W < 0) = 2^n - 1 - n = 2^n - (n+1)$$

Consequently, the total number of reversible solutions with a net input or output of work is

$$Z_{rev}(W > 0) + Z_{rev}(W < 0) = 2^{n+1} - 2(n+1)$$

Example 1.1
A heat pump transformer with four temperature levels is evaluated. Figure 1.16 shows a schematic drawing. Heat is provided at temperature T_3, and waste heat is rejected at temperature T_0. The amount of driving heat Q_3 and waste heat Q_0 is the same. Heat is pumped from T_1 to T_2. Find the heat ratio Q_2/Q_3. It is assumed that the temperatures have the following values: $T_3 = 160°C$, $T_2 = 100°C$, $T_1 = 80°C$, and $T_0 = 30°C$.

Solution: The First Law for the heat pump transformer reads:

$$-Q_0 + Q_1 - Q_2 + Q_3 = 0$$

and the Second Law:

$$-\frac{Q_0}{T_0} + \frac{Q_1}{T_1} - \frac{Q_2}{T_2} + \frac{Q_3}{T_3} \leq 0$$

By eliminating Q_1 and considering that the amounts of Q_3 and Q_0 are equal, the following expression is obtained:

$$Q_3\left(\frac{1}{T_3} - \frac{1}{T_0}\right) + Q_2\left(\frac{1}{T_1} - \frac{1}{T_2}\right) \leq 0$$

When the heat pump transformer operates reversibly, the following expression is obtained for the heat ratio:

$$\frac{Q_2}{Q_3} = \left(\frac{\dfrac{1}{T_2} - \dfrac{1}{T_1}}{\dfrac{1}{T_3} - \dfrac{1}{T_0}}\right)^{-1}$$

For the specified temperatures, the heat ratio amounts to 6.5.

REFERENCES

1. Nesselmann, *Zeit. ges. Kälteindustne*, Vol. 41, pg. 73, (1934).
2. Nesselmann, *Zeit. ges. Kälteindustne*, Vol. 42, pg. 8, (1935).

CHAPTER TWO

Working Fluids

The efficiency of reversible cycles is independent of the properties of the working fluids used. However, the efficiency of a realistic cycle is determined to a large degree by the properties of the working fluids which cause various irreversibilities. The first cost of any energy conversion apparatus is definitely dependent on the working fluid properties. This chapter describes the properties of several typical working fluid mixtures and their applications. Typical diagrams of fluid properties are discussed.

2.1 WORKING FLUIDS IN THERMODYNAMIC DIAGRAMS

When designing a heat pump, the most important variables to be considered are pressure, p, temperature, t, composition, x, enthalpy, h, specific volume, v, entropy, s, and transport properties. To display all variables, a multidimensional diagram is required, which is not practical. Several two-dimensional diagrams are in use. These diagrams show any two variables on their axes and display other variables as sets of curves of constant properties such as isobars, isotherms, etc.

Usually, T,s-, ln(P),h-, or h,s-diagrams are used for a quantitative evaluation of the coefficient of performance of vapor compression and expansion processes which use pure working fluids. However, when mixtures are used, in absorption processes for example, the additional variable, composition, must be considered. Historically, enthalpy-composition diagrams were preferred, with temperature and pressure as parameters. Figures 2.1 and 2.2 are examples.

For the sake of simplicity, this explanation focuses on two-component mixtures only. From a thermodynamic point of view, a two-component mixture possesses an additional degree of freedom as compared to a pure fluid. The new variable, the composition, is defined as:

$$x = \frac{\text{mass of one component [kg]}}{\text{total mass of all components [kg]}}$$

2.1.1 Temperature-Concentration Diagram

When the liquid and vapor phase of a mixture coexist, the saturation temperature varies with the concentration, even though the pressure is constant. This is not possible when a pure fluid is used. Figure 2.3 shows the schematic temperature-concentration diagram (T,x diagram) for a mixture of two components, A and B, at constant pressure. The mixture-concentration axis ranges from 0 (only pure A is present) to 1.0 (only component B is present). The area below the boiling point line represents subcooled liquid. The area above the dew point line represents superheated vapor. The area enclosed by the boiling and dew point lines is the two-phase range. The boiling point for a mixture of the concentration x is located at the intersection of the isostere (line of constant concentration) with the boiling point line. The boiling point line indicates the temperature at which the first vapor bubble is formed for the specified pressure and concentration. The boiling points of the pure components, T_A and T_B, are found on the respective ordinates. The boiling point of component A is higher than that of component B. The dew point line indicates the temperature at which the first liquid droplet is formed when a gas mixture of a given concentration is cooled.

To demonstrate the use of the diagram, an evaporation process at constant pressure is discussed as an example. The process begins with subcooled liquid at point 1″. Points with a

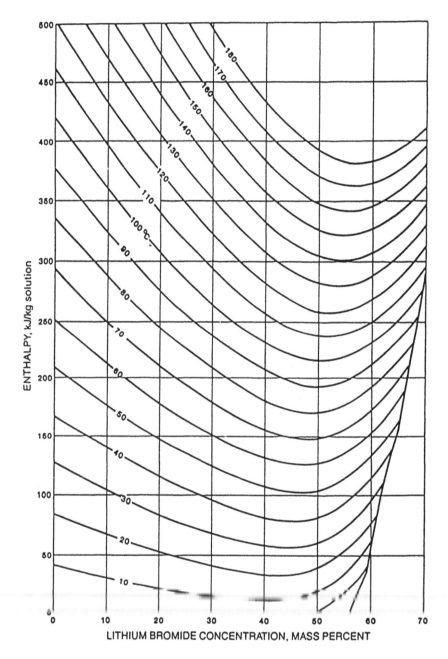

Figure 2.1 Enthalpy-concentration diagram for lithium bromide/water solutions. Only the liquid region is shown. (From *ASHRAE Handbook of Fundamentals,* American Society of Heating Refrigeration and Air-Conditioning Engineers, Atlanta, 1989. With permission.)

double prime indicate liquid phase; points with a single prime indicate vapor phase. When heat is added, the temperature increases, and the boiling point line is reached. This is point 2. Here the first vapor bubble forms. The concentration of the vapor in the bubble which is in thermal equilibrium with the surrounding liquid is found at point 2'. The vapor is enriched in component B which, at the same temperature, has a higher vapor pressure than that of A. As more heat is added, the evaporation process continues to point 3. If the entire vapor remains in contact with the remaining liquid, then the concentration of the vapor in equilibrium with the remaining liquid is represented by point 3'. The concentration of the remaining liquid is

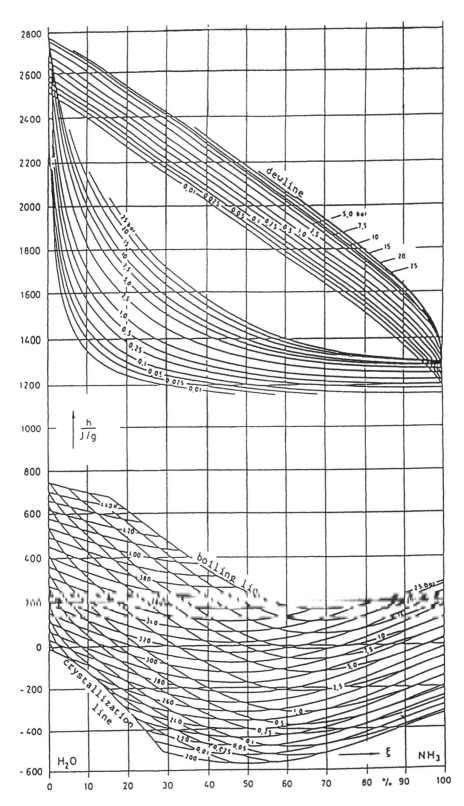

Figure 2.2 Enthalpy-concentration diagram for ammonia/water solutions. (From *Kältemaschinenregeln*, 7th Ed., Verlag C. F. Müller, Karlsruhe, Germany, 1981. With permission.)

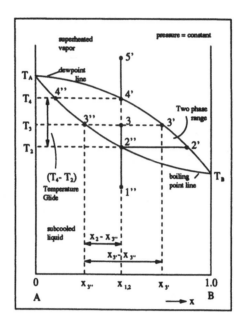

Figure 2.3 Temperature-concentration diagram of a zeotropic mixture.

indicated by 3″. At this point, liquid has been depleted of component B. The vapor contains more of component A than previously. As the evaporation process proceeds, the concentrations of the liquid and vapor phases follow the boiling and dew point lines. When point 4′ is reached, the evaporation process is completed. The vapor has the same concentration as the original subcooled liquid, and the concentration of the last evaporated droplet is indicated by point 4″. Further addition of heat produces superheated vapor at point 5. During the constant pressure evaporation process, the saturation temperature changed from T_2 to T_4. The temperature difference (T_4-T_2) is termed "temperature glide".

Vapor quality x_q (defined as the ratio of mass of vapor over total mass) at point 3 can be calculated based on a mass balance for the mixture and one pure component. The calculation is based on 1 kg of total mass.

$$m_{3''} + m_{3'} = 1 \tag{2.1}$$

$$m_{3''} \, x_{3''} + m_{3'} \, x_{3'} = 1 \, x_2 \tag{2.2}$$

After elimination of $m_{3''}$, the vapor quality is obtained as:

$$x_q = m_{3'} = \left(x_2 - x_{3''}\right)/\left(x_{3'} - x_{3''}\right) \tag{2.3}$$

The vapor quality is represented in Figure 2.3 by the ratio of the distance $(x_2 - x_{3''})/ (x_{3'} - x_{3''})$ and can be expressed as a function of concentrations only.

As a general rule, the temperature glide increases with increasing difference between the boiling points of the two pure components.

The mixture in Figure 2.3 is traditionally termed a "non-azeotropic mixture" or, in more recent literature, "zeotropic mixture". The name implies that, in phase equilibrium, the concentrations of the vapor and liquid phases are always different. Some fluids form azeotropic mixtures, for example a mixture of R12 and R152a or a mixture of water and ethanol. For an azeotropic mixture, the concentration of the liquid and vapor phase is identical at a certain concentration, as shown in Figure 2.4. At this concentration, the temperature glide is zero.

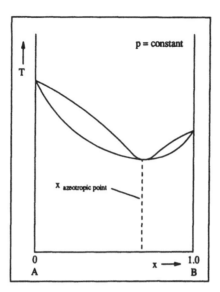

Figure 2.4 Temperature-concentration diagram of an azeotropic mixture. The boiling point of the azeotrope is lower than those of the pure components.

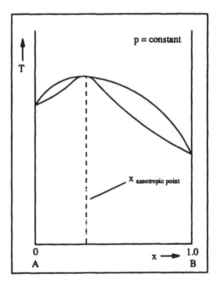

Figure 2.5 Temperature-concentration diagram of an azeotropic mixture. The boiling point at the azeotrope is higher than those of the two pure components.

At all other concentrations, the mixture exhibits zeotropic behavior. The difference in concentration between the liquid and vapor phases changes its sign when the overall concentration varies from a value less than the azeotropic concentration to a value larger than the azeotropic concentration. There are two types of azeotropes. These are distinguished from one another by the location of the boiling point at the azeotropic concentration relative to the boiling points of the pure fluids. The boiling point can either be higher than the boiling point of any of the two constituents of the mixture or lower than the boiling point of any of the two constituents (Figure 2.5).

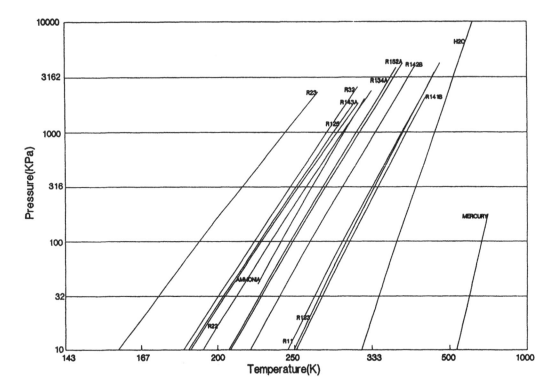

Figure 2.6   Vapor pressure curves for selected refrigerants in a pressure-temperature diagram.

2.1.2   Pressure-Temperature Diagram

For preliminary investigation and comparison of working fluids, the pressure-temperature diagram [ln(P),(-1)/T diagram] can be used. Figure 2.6 shows vapor-pressure curves for a number of refrigerants. For this particular choice of coordinates, ln(P), (-1)/T, the vapor-pressure curves become almost straight lines reaching from the triple point to the critical point. The area to the left of the vapor-pressure curve shows higher pressures and lower temperatures than saturation values and represents subcooled liquid. The area to the right shows, in contrast, higher temperatures and lower pressures and represents superheated vapor. The two-phase range is represented by the vapor-pressure line itself. Azeotropic mixtures yield similar vapor-pressure lines as, shown in Figure 2.6 (for example, R502, a mixture of R22 and R115).

For mixtures of fluids, curves similar to vapor-pressure curves can be found by plotting the vapor pressure vs. (-1)/T for isosteres of the saturated liquid phase. Figure 2.7 shows, as an example, ammonia/water. The space between the vapor-pressure curves of the pure constituents of the mixture is termed "solution field". The pressure-temperature diagrams of solutions are later used extensively for the representation of cycle configurations.

2.1.3   Pressure-Enthalpy Diagrams

Figures 2.8 and 2.9 display pressure enthalpy diagrams (ln(P),h diagrams) for two pure fluids, ammonia and R22. Traditionally, these diagrams have been used in the refrigeration field for the evaluation of vapor-compression cycles. Figure 2.10 shows a schematic diagram of the ln(P),h. The subcooled liquid region, the superheated vapor region, and the two-phase region are marked. The critical point is indicated, and a typical isotherm is shown. It is considered an advantage of the ln(P),h-diagram that the amounts of heat and work exchanged in energy conversion processes are represented as distances within the diagram. Figure 2.11 shows a ln(P),h diagram for a zeotropic mixture of R22 and R142b at a concentration of 50% by mass. In the single-phase regions, the diagram is very similar to that of a pure fluid. In the two-phase

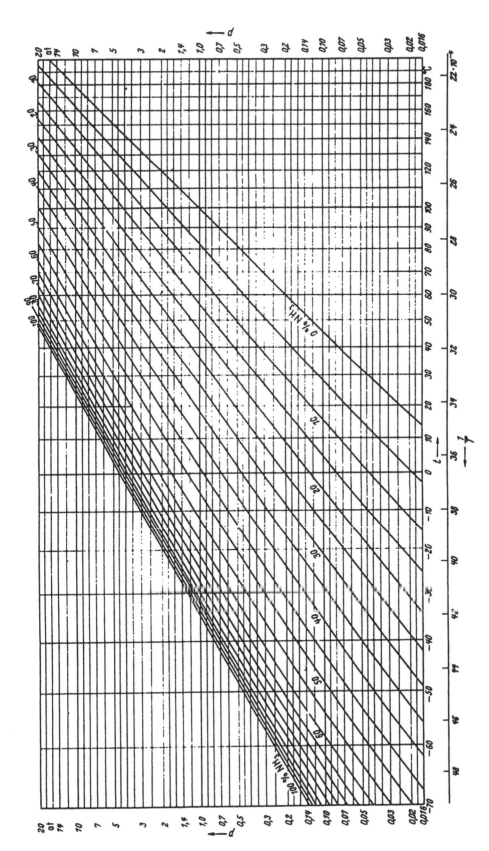

Figure 2.7 ln(P),(−1)/t diagram for ammonia/water solutions.

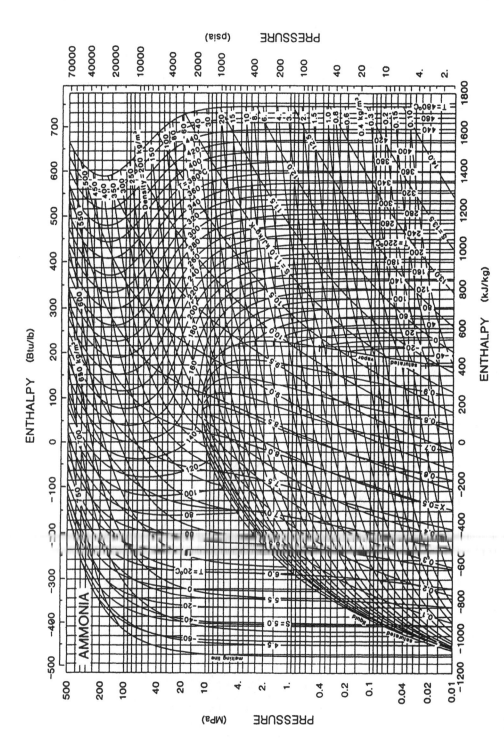

Figure 2.8 Pressure-enthalpy diagram for ammonia. (From *ASHRAE Handbook of Fundamentals*, American Society of Heating Refrigeration and Air-Conditioning Engineers, Atlanta, 1989. With permission.)

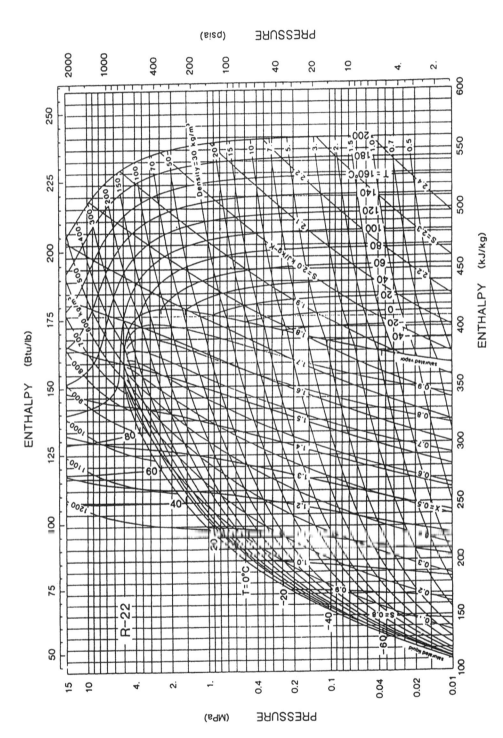

Figure 2.9 Pressure-enthalpy diagram for refrigerant R22. (From *ASHRAE Handbook of Fundamentals*, American Society of Heating Refrigeration and Air-Conditioning Engineers, Atlanta, 1989. With permission.)

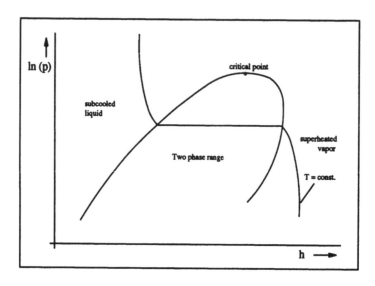

Figure 2.10   Schematic pressure-enthalpy diagram.

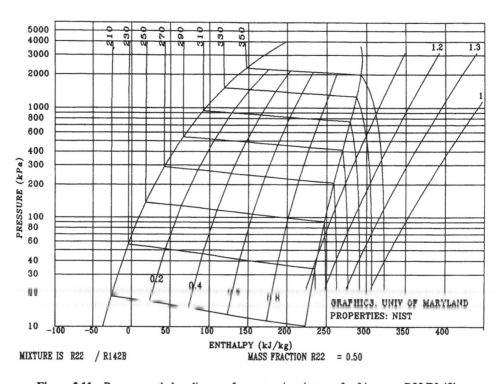

MIXTURE IS R22 / R142B   MASS FRACTION R22 = 0.50

Figure 2.11   Pressure-enthalpy diagram for zeotropic mixture of refrigerants R22/RL42b.

region, a significant difference can be observed. The isotherms are sloped toward lower pressures as the enthalpy increases. Thus, the temperature glide for a constant pressure evaporation process is displayed. An isobar, crossing the two-phase range, intersects at the saturated vapor line, an isotherm of a temperature that is higher than the temperature of the isotherm at the intersect of the isobar at the saturated liquid line. The general shape of the isotherm in the two-phase region can be concave, convex, or s-shaped.

ln(P),h diagrams for mixtures are valid only for the specified overall concentration. When the concentration is changed, a new diagram must be generated.

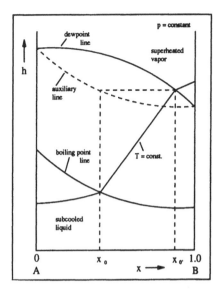

Figure 2.12   Schematic enthalpy-concentration diagram.

2.1.4 The Enthalpy-Concentration Diagram

The enthalpy-composition diagram was originally introduced by Mollier and Bosnjakovic. It found wide applications for the design of absorption heat pumps and distillation equipment. It provides information about enthalpies, composition of the liquid and vapor phase, temperature, and pressure. It is, therefore, convenient to use.

The enthalpy-composition diagram (h,x diagram), as shown in Figures 2.12 and 2.2, is divided into three regions. The lower region displays properties of the subcooled liquid. It has a set of isobars and isotherms which are not shown in Figure 2.13 for the sake of simplicity. These can be seen in Figure 2.2. Although the isobars and isotherms represent saturated liquid, they are also valid for subcooled liquid, assuming that the enthalpy of the liquid phase is independent of the pressure.

The middle of the diagram, Figure 2.12, represents the two-phase region. This area is bordered by the boiling point line and the dew point line. These two lines are isobars. For increasing pressure, the sets of isobars shift two higher enthalpy values. The distance at the end points represents the latent heat of evaporation of the pure fluids. The isotherms are straight lines between the boiling point line and the dew point line. A typical isotherm is shown in Figure 2.12. Each pair of dew and boiling point lines has its own set of isotherms.

The upper region of the diagram is the superheated vapor area. In this area, the isotherms are usually almost straight lines, since the heat of mixing of gases is negligible. Again, each dew point line has its own set of isotherms. They are usually not displayed, to avoid overcrowding. The isobars in the liquid region represent the boiling point line.

In Figures 2.12 and 2.2, the vapor concentration can be determined by finding the intersection of the liquid isostere x_0 with the auxiliary line in the two-phase region. From there, one follows a line of constant enthalpy (which is the enthalpy of the vapor) to the dew point line. The intersection with the dew point line indicates the vapor concentration $x_{0'}$. This procedure is shown in Figure 2.13. In Figure 2.1, the isotherms in the two-phase region are not shown. Their end point on the dew line can be constructed with the help of the auxiliary lines.

2.2 PROPERTIES OF WORKING FLUIDS

New developments for heat conversion processes are anticipated when multicomponent working and/or absorber fluids are utilized. There exists the potential for better efficiencies

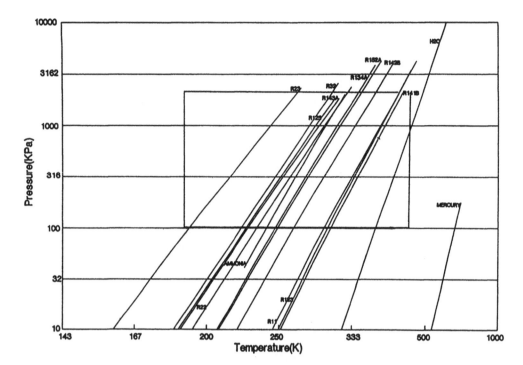

Figure 2.13 Vapor pressure curves for selected refrigerants. The window indicates the temperature and pressure range of interest.

and new variations of configurations. The following outline shows the possibilities and the assessment for research focusing on new working fluid mixtures. In general, the working fluids and their mixture should have the following properties:

A. The ratio of the heat capacity over latent heat should be as small as possible to obtain a high efficiency. In contrast to popular opinion, it is the absolute value of the heat of evaporation per unit mass, a rather misleading measure for the quality of a fluid. Molecules composed of many heavy atoms, such as halogenated hydrocarbons, have high heat capacities compared to the latent heat. They are not desirable from this point of view.

B. The volumetric capacity should be as high as possible. Since the latent heat per mole divided by the specific volume per mole is for a given temperature and pressure range, almost fluid independent, the volumetric capacity is essentially fluid independent and proportional to the pressure.

C. The pressure ratio should be as small as possible to obtain acceptable operating conditions for compressors and solution pumps.

D. The toxicity of the fluid mixture should be as low as possible to maximize the safety of personnel working with the equipment during manufacture and maintenance and of the general public in case of an accidental release of working fluid.

E. The fluids should be harmless to the environment. The fluid should not contribute to global warming, ozone depletion, or production of harmful derivatives or radicals.

F. For the safety of any equipment as well as of manufacturing and service personnel, it is desirable that fluids not be flammable or explosive.

G. In order to obtain reasonable equipment life, fluids used should be noncorrosive. When using fluids such as lithium bromide/water and ammonia/water, it is possible to add corrosion inhibitors which reduce or eliminate corrosion.

H. The fluids should be thermally stabile and not decompose at normal operating temperatures. The thermal stability and corrosive properties limit the temperature range in which energy conversion systems operate.

I. The fluids should be compatible with other materials necessary for the construction and operation of energy conversion equipment such as container materials, seals, and sensors, and fluids such as compressor oils, bearing lubricants, cleaning solvents, heat transfer additives, and impurities.

2.2.1 Working Fluids for Vapor-Compression Systems

In vapor-compression systems, halogenated hydrocarbons are used almost exclusively. A window in terms of pressures and temperatures is shown in Figure 2.13. It frames the vapor pressures of acceptable working fluids. While the temperature is determined by the respective tasks, the pressure range is prescribed by the requirements and type of the compressors (reciprocal, screw, or centrifugal compressor). High pressures require more effort. In case the pressure drops below a certain minimum, the compressor becomes too large to be practical because of the low vapor density. (Exceptions are R11 compressors and water-vapor compressors for ice production). Whether a working fluid is located within the window is determined by the latent heat of evaporation per mole (not per unit of mass) and its boiling point. In new developments, mixtures of working fluids are used to increase efficiency by taking advantage of the gliding temperatures. When changing the composition of the circulating mixture, the capacity can be adjusted to match, for example, a changing heating load due to varying outdoor temperatures.

2.2.2 Working Fluids for Absorption Systems

For absorption heat pumps, a suitable absorber has to be found, in addition to the refrigerant. The following pairs are frequently employed or discussed:

NH_3/H_2O

H_2O/H_2O-LiBr solution

CH_3OH/CH_3OH-salt solution

R22 ($CHClF_2$)/E181 or other organic solvent

R133a (CH_2ClCF_3)/ETFE

Similar to vapor-compression heat pumps, the refrigerant for absorption heat pumps is determined by the selected temperature range and the requirement that the pressure levels have to be technically feasible. A window similar to the one in Figure 2.13 is shown in Figure 2.14, showing the location of acceptable refrigerants. Trouton's rule (common intersection of vapor pressure lines for T approaching infinity) determines whether or not a working fluid is located in the technically acceptable range. In addition, all of the criteria discussed in Chapter 2.2 are of concern for the selection of the working fluid and the absorbent. Three characteristics are added here specifically for absorption equipment:

A. The absorbent should exhibit a low vapor pressure compared to the working fluid. Its vapor pressure curve should be located as far right in Figure 2.14 as possible in relation

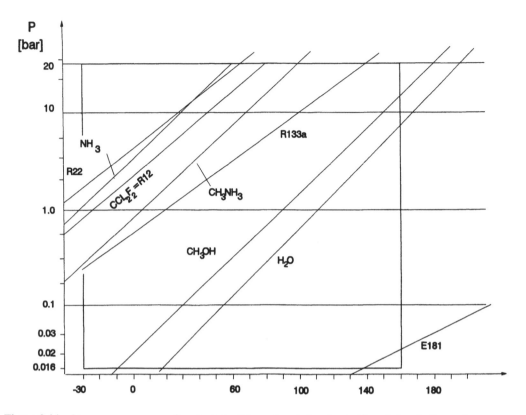

Figure 2.14 Vapor pressure curves for selected refrigerants and absorbents. The window indicates the temperature and pressure range of interest.

to working fluid. The vapor pressure should be determined only by the working fluid itself.

B. The heat capacity of the refrigerant and absorbent divided by the latent heat of the refrigerant should be as low as possible.

C. Additives for the enhancement of heat and/or mass transfer should be available without hampering any of the other characteristics such as corrosiveness, thermal stability, and toxicity.

When the working pair NH_3/H_2O seen in Figure 2.8 is used in heat pumping with condenser temperatures of about 50°C and evaporator temperatures of 0°C, the pressure levels are 4 bar and 25 bar, respectively. While these pressures are acceptable, the toxicity of NH_3 requires strict observance of safety precautions during installation and maintenance. The pressures of the working pair $H_2O/LiBr$, Figure 2.15, vary between 8 mbar and 200 mbar for single-effect units. The main disadvantage of this system for heat pump applications is the narrow solution field, which is limited by crystallization. Absorber temperatures are limited to about 40°C. The limitation of freezing water in the evaporator is not severe, since the freezing point can be lowered by the addition of a small amount of LiBr. The vapor pressure reduction for the new freezing point caused by the added salt is determined by the ratio of heat of fusion to heat of vaporization. When water is used, the ratio is small: 1:7.

This working fluid pair is frequently applied in water chillers. The units have to be designed so that all tubing and valves for vapor are avoided because of the low density of the steam. This is described in more detail in Chapter Four.

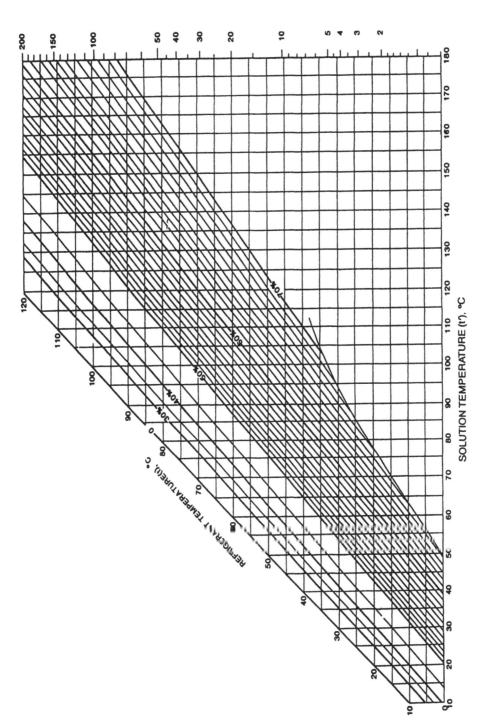

Figure 2.15 Solution field for lithium bromide/water.

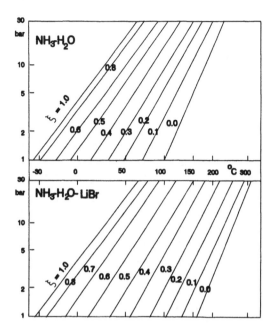

Figure 2.16 Comparison of solution fields for ammonia/water and ammonia/water/lithium bromide.

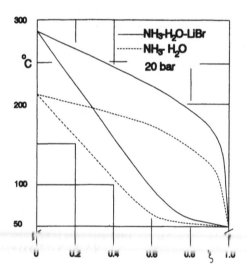

Figure 2.17 Comparison of temperature-concentration diagrams for ammonia/water and ammonia/water/lithium bromide.

In recent years, binary absorber fluids have been studied. For example, an aqueous solution of LiBr can be used as an absorber for NH_3 or for CH_3NH_2. The solution field is extended towards significantly higher temperatures, Figures 2.16 to 2.18. For a given desorber temperature and pressure, the solution has a higher ammonia (or methylamine) composition in relation to the H_2O content in the liquid phase. This reduces the rectification requirements significantly. A comparison of Figures 2.16 to 2.18 illustrates the principle.

Similar improvements can, in principle, be obtained for H_2O/LiBr with H_2O as the working fluid. The crystallization limit is moved to higher temperatures by adding a second liquid absorbent, glycol (Figure 2.19). The plain system H_2O/glycol cannot be used because of the high vapor pressure of glycol. However, added LiBr reduces the vapor pressure similar to NH_3/H_2O, so that this system becomes an acceptable mixture improved by the extension of

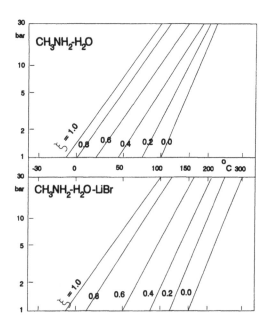

Figure 2.18 Comparison of solution fields for methylamine/water and methylamine/water/lithium bromide.

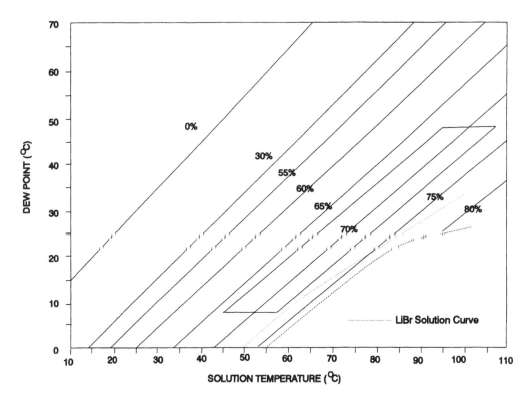

Figure 2.19 Solution field for water/glycol/lithium bromide.

the crystallization limit. In addition, the use of binary working fluids, which must have a common absorbent, can lead to improved efficiencies. This follows the same principle that applies to vapor-compression heat pumps. Examples are $NH_3+CH_3NH_2$/aqueous solution of LiBr and H_2O-CH_3OH/solution of H_2O-CH_3OH-LiBr-ZnBr.

The working fluid combinations described so far are the most commonly used. However, recent research efforts resulted in additional fluids which can be used particularly for absorption systems. A comprehensive reference list for fluid properties is given at the end of this chapter.

2.3 RESEARCH AREAS FOR NEW FLUIDS

An improvement of today's most common working fluids or the replacement with better ones is desirable. The following research areas for new fluids can be identified:

A. New (single-component) working fluids for vapor compression and expansion processes are needed, especially high temperature fluids.

B. Azeotropic or near-azeotropic mixtures are desirable to obtain fluids in pressure ranges where pure components are not available.

C. Zeotropic mixtures are needed for performance improvement.

D. New working fluid combinations for absorption heat pumps are needed, especially for high temperature applications.

E. Azeotropic and zeotropic refrigerant mixtures for a common absorbent can be of interest for performance improvements.

F. New working fluid combinations for absorption heat pumps with a non-condensible inert gas are desirable.

G. New working fluid combinations for absorption heat pumps with a condensing inert gas and good heat transfer properties are of interest.

H. New working fluid combinations for absorption heat pumps with mixing gaps may offer interesting new cycle designs and energy-saving options.

I. New working fluid combinations for solid-vapor systems are required.

REFERENCES

1. Niebergall, W., Arbeitsstoffpaare für Absorptions-Kälteanlagen and Absorptions-Kühlschränke, Mühlhausen/ Thür. Verlag für Fachliteratur Richard Markewitz, 1949.

2. *ASHRAE Handbook of Fundamentals*, American Society of Heating Refrigeration and Air-Conditioning Engineers, Atlanta, 1989.

3. DKV (Hrsg.), Kältemaschinenregeln, C. F. Müller Verlag, Karlsruhe, 7. Aufl., 1981.

4. Ziegler, B., Wärmetransformation durch einstufige Sorptionsprozesse mit dem Stoffpaar Ammoniak-Wasser, Diss. ETH Nr. 7070, Zürich, 1982.

5. Ziegler, B., Trepp, C., Equation of state for ammonia/water mixtures, *Int. J. Refrigeration*, Vol. 7, pp 101–107, 1984.

6. Macriss, R. A., Gutraj, J. M., Zawacki, T. S., Absorption Fluids Data Survey: Final Report on Worldwide Data, U.S. Department of Energy, Office of Building and Community Systems, ORNL/Sub/84-47989/3.

7. Niebergall, W., Sorptions Kältemaschiren Handbook der Kültetechmik Bd. 7, Springer Verlag, Berlin, 1981.

CHAPTER THREE

Vapor-Compression Heat Pumps and Power Plants

In conventional Rankine Cycles, phase change processes are described as "evaporation" or "condensation". These terms usually imply that a pure working fluid is used and that the phase change process is complete. At the evaporator outlet and at the condenser inlet, the vapor quality is one. This chapter describes generalized Rankine Cycles which use working fluid mixtures. The phase transition gas-liquid and vice versa are more precisely described by the terms "desorption" and "absorption" instead of "evaporation" and "condensation". Examples are vapor-compression systems with solution circuits and absorption heat pumps and heat transformers. Emphasis is placed on the graphic representation of these cycles.

The most commonly employed thermodynamic cycle in heat conversion systems is the Rankine Cycle or, more generally, the Rankine Sorption Cycle. This class of cycles is distinguished from gas cycles by the two-phase changes which occur during the course of the cycle. For heat engines, the pressure increase is accomplished in the condensed phase. The expansion occurs in the gas phase. The Rankine Cycle has the advantage that the net output of work is essentially identical to the expansion work. This is not the case in gas cycles. For vapor-compression heat pumps, the expansion occurs in the liquid phase. Since the amount of expansion work which could be recovered is small compared to the compression work, it is possible to replace the expansion turbine by a throttle. The heat-conversion processes in absorption heat pumps are also accomplished in total or in part by Rankine Sorption Cycles.

3.1 THE RANKINE CYCLE WITH INPUT AND OUTPUT OF WORK

Figure 3.1 shows in a three dimensional P,ln(v),T representation the well-known Rankine Cycle, 1–2–3–4–5–6, as it is employed in steam power plants and Rankine Cycles working with organic fluids. State points 6–5–4–3–7 denote the Rankine Cycle as used in vapor-compression heat pumps. The arrows within the diagram have the following meaning: If an arrow is pointing towards a line which represents a process line, then heat is supplied to the working fluid, and vice versa. In contrast to the gas cycles of Figure 3.1 (Stirling Cycle, Joule Brayton Cycle, and Carnot Cycle), the Rankine Cycle shows two-phase changes. The necessary increase in pressure occurs in the condensed phase of the working fluid, and only a minimum amount of work is required. Therefore, the work gained by the expansion 3–6 represents approximately the net work produced. For gas cycles, the net work is the difference between two large numbers and is strongly dependent on the efficiencies for the expansion and compression machines. When the Rankine Cycle is used for heat pumping and when temperatures and pressures are far below the critical point, the expansion 3–2–1 can be accomplished by a simple throttle (3–7) causing relatively small losses of work. Figures 3.2 (a-g) and 3.3 (a-g) show projections of the Rankine Cycles with input and output of work in various diagrams. The representation ln(P) vs. -1/T, which is a projection parallel to the ln(v) axis of Figure 3.1, is included here for comparison because this diagram is later referred to extensively. Figures 3.2h and 3.3h show the essential components of the respective cycles. The small expander in Figure 3.3h is usually replaced by a throttle device. The T, s diagram in Figure 3.2d is traditionally used to design steam-based power generation cycles. Figure 3.4 shows the diagram in a larger scale for water. The ln(P),h diagram for Figure 3.2e is

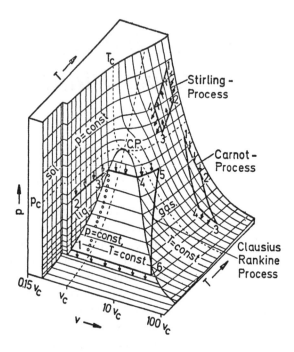

Figure 3.1   The Rankine Cycle and various gas cycles in a P,ln(v),T diagram.

traditionally used for the design of vapor-compression heat pumps. Figures 2.9 and 2.10 (Chapter Two) show examples for ammonia and refrigerant R22. Usually, the diagrams of Figures 3.2d and e contain the same information, and both are useful in designing power generation or heat pump cycles.

3.2   VAPOR-COMPRESSION HEAT PUMPS, REFRIGERATION UNITS, AND HEAT ENGINES

Figure 3.5 shows schematically the essential components of a vapor-compression heat pump (counter-clockwise circulation of the working fluid) and a heat engine (clockwise circulation of the working fluid) in a pressure vs. temperature diagram. The location of the heat exchangers accommodating a phase change indicate pressure and temperature levels relative to each other. The shift of the two components relative to each other is determined by the slope of the vapor pressure line if the axes are chosen to be ln (P) vs. (−1)/T. For the compressor K, Figure 3.5 indicates only the high and low side pressures correctly. Temperatures are not indicated by the location of this component. In heat pumping, component A represents an evaporator, component B represents a condenser, component K represents a compressor, and component D represents a throttle. Component P is only present when a solution circuit is included. The closer the temperature T_1 approaches the critical temperature, the larger is the ratio of specific heat over latent heat of evaporation, the higher the gains in efficiency when the throttle is replaced by a turbine. When used as a power plant, component B represents the boiler, component K represents the turbine, component A represents the condenser, and component P represents the feed pump. Component D is not present. In case of a power generation cycle, the efficiency η is defined as: η = (net work output)/(heat input). The efficiency of a vapor-compression heat pump is defined as: η_{HP} = (heat output) / (net work input).

Example 3.1: Steam Power Plant (Heat Engine)
The performance of a steam power plant is evaluated. The expansion process in the turbine is assumed to be isentropic. There is no pressure drop in the

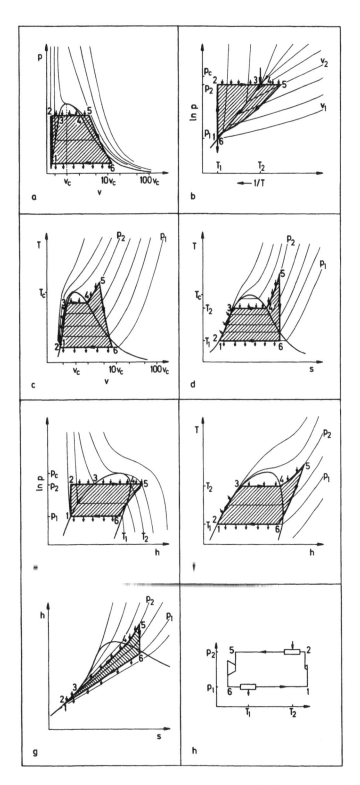

Figure 3.2 Projections of the work-producing Rankine Cycle on various planes.

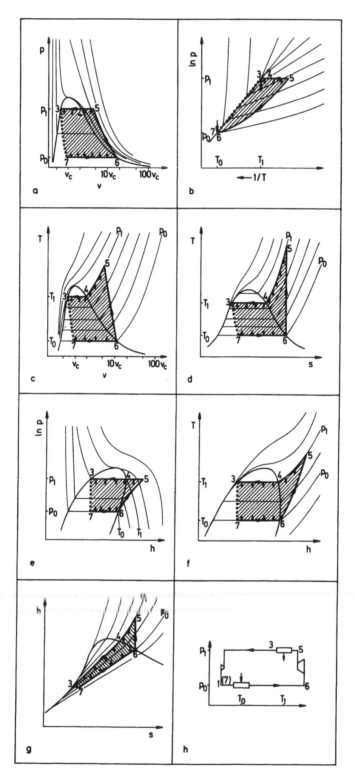

Figure 3.3   Projections of the work-consuming Rankine Cycle on various planes.

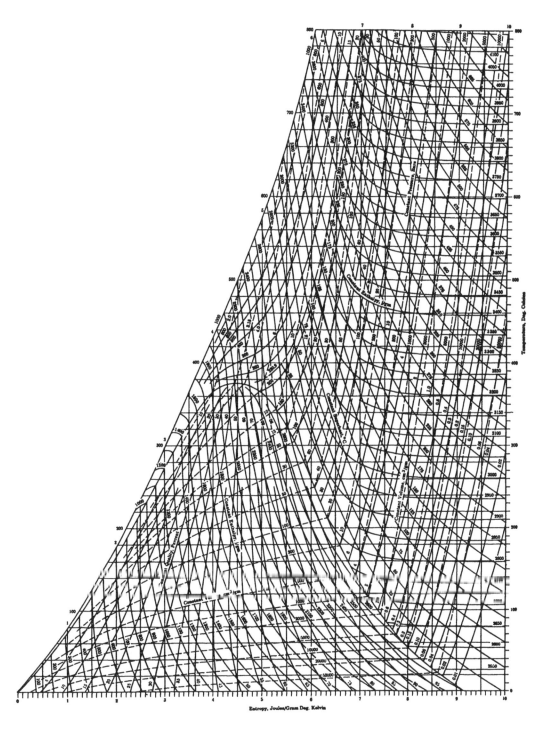

Figure 3.4 Temperature-entropy diagram for water. (From Keenan, J. H., Keyes, F. G., Hill, D. G., and Moore, J. G., *Steam Tables*, John Wiley & Sons, New York, 1978. Reprinted with permission of the American Society of Heating, Refrigerating, and Air Conditioning Engineers, M. A. Keenan, and John Wiley & Sons.)

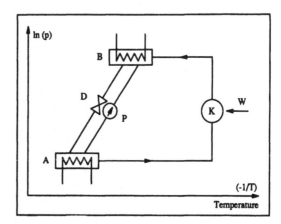

Figure 3.5 Schematic of vapor-compression heat pump with solution circuit.

boiler and condenser, and the efficiency of the liquid feed pump is assumed to be 1.0. The temperature of the saturated steam entering the superheater is 260°C, the temperature of the steam leaving the superheater is 360°C, and the temperature of the condensing steam is 40°C.

(a) What is the amount of the heat supplied to the boiler and superheater per kilogram of steam?

(b) What is the amount of the heat rejected at the condenser per kilogram of steam?

(c) What is the amount of work produced by the turbine and consumed by the liquid feed pump per kilogram of steam?

(d) Using the above answers, what is the efficiency of the power plant?

Solution: To obtain the thermodynamic data, the T,s diagram of Figure 3.4 is used. Figure 3.6 shows a schematic diagram of the T,s diagram. State point 1 is the condenser outlet, 2 is the liquid feed pump outlet, 3 the boiler outlet, 4 the superheater outlet, and 5 the turbine outlet. All property values are listed in Table 3.1.

Table 3.1   Thermodynamic Properties of the State Points

	T(C)	P(kPa)	h(kJ/kg)	s(kJ/kg/K)
1	40	7.4	180	0.55
2	41	4690	185	0.55
3	260	4690	2797	6.00
4	360	4690	3078	6.50
5	40	7.4	2015	6.50

The evaluation is based on 1 kg of working fluid. The enthalpy of the liquid stream after the pump is found by adding the pump work to the enthalpy at the pump inlet. The pump work amounts to $w_p = v(P_2 - P_1)$, with v the specific volume of the liquid water and $P_{1,2}$ the pressure at the respective state points. It is assumed that the water is incompressible and that the compression

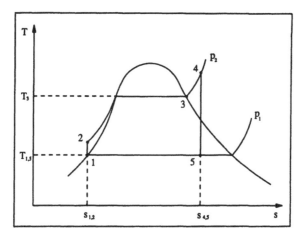

Figure 3.6 Schematic temperature-entropy diagram.

process is isentropic. The temperature of the water is increased by about 0.8 K. The enthalpy h_5 at the turbine outlet is found at the intersection of the isentrope through state point 4 with the isobar P_1 in the two-phase range. The vapor quality at this point is 0.77.

The boiler heat q_b, the superheat q_s, and the condenser heat q_c amount to:

(a) $q_b = h_3 - h_2 = 2612 \text{ kJ/kg}$

$q_s = h_4 - h_3 = 281 \text{ kJ/kg}$

(b) $q_c = h_5 - h_1 = 1835 \text{ kJ/kg}$

The work produced by the turbine and consumed by the pump amounts to:

(c) $w_t = h_4 - h_5 = 1063 \text{ kJ/kg}$

$w_p = 4.7 \text{ kJ/kg}$

The efficiency η is then

(d) $\eta = \left(w_t - w_p\right)/\left(q_b + q_s\right) = 0.37$

Remarks: Because the work input to the liquid feed pump is small compared to the work produced by the turbine, it is frequently neglected. The turbine outlet is saturated with a vapor quality of less than 1.0. This is usually avoided in actual machinery to reduce erosion to the turbine blades by droplets of liquid. To avoid a saturated turbine outlet, the superheat temperature has to be raised or the high side pressure level reduced. The results can be checked for consistency when an energy balance is applied to the entire cycle:

$$q_b + q_s + w_p - q_c - w_t = 0$$

$$2612 \text{ kJ/kg} + 281 \text{ kJ/kg} + 5 \text{ kJ/kg} - 1835 \text{ kJ/kg} - 1063 \text{ kJ/kg} = 0$$

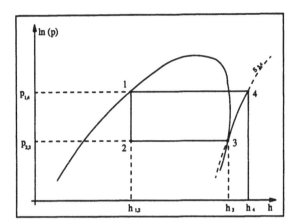

Figure 3.7   Schematic pressure-enthalpy diagram with a superimposed vapor-compression cycle.

The fact that the energy balance is fulfilled indicates only that the enthalpy values were applied consistently in the individual equations. It does not confirm that the enthalpy values were correctly obtained from the thermodynamic charts or tables.

Example 3.2: Vapor-Compression Heat Pump

A vapor-compression heat pump using ammonia as the working fluid is evaluated. The temperature of the evaporating refrigerant is $-10°C$, while the temperature of the condensing refrigerant is $+40°C$. An idealized Rankine Cycle with isentropic compression, no pressure drop in the heat exchangers, and connecting lines and isenthalpic expansion process is assumed.

(a) What is the amount of the condenser heat, q_c, the evaporator heat, q_e, and the compressor work w_{com} per 1 kg of refrigerant?

(b) What is the COP (coefficient of performance) based on the above results?

(c) How large is the pressure ratio?

(d) How large is the volumetric capacity?

(e) What is the value of the temperature of the vapor at the compressor outlet?

Solution: To obtain all necessary thermodynamic data, the $\ln(P),h$ diagram of Figure 2.8 (Chapter Two) is used. Figure 3.7 shows a schematic diagram. State point 1 is the condenser outlet, the evaporator inlet is 2, the evaporator outlet is 3, and the compressor outlet is 4. The evaluation is based on 1 kg of refrigerant.

The state properties for the evaporator inlet are found where the line of constant enthalpy through the condenser outlet, point 1, intersects with the low pressure isobar. State point 4 is located at the intersection of the isentrope through point 3 with the high pressure isobar. The properties are listed in Table 3.2.

(a) The condenser heat, evaporator heat, and compressor work amount to:

Table 3.2 Thermodynamic Properties of the State Points

	T(C)	P(kPa)	h(kJ/kg)	s(kJ/kg/K)
1	40	1560	−572	6.39
2	−10	290	−572	6.45
3	−10	290	488	10.45
4	118	1560	745	10.45

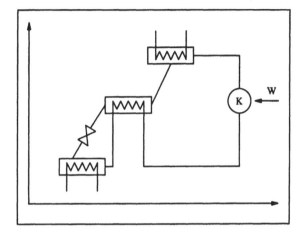

Figure 3.8 Heat exchange between liquid and vapor streams of the working fluid.

$$q_c = h_4 - h_1 = 1317 \text{ kJ/kg}$$

$$q_e = h_3 - h_2 = 1060 \text{ kJ/kg}$$

$$w_{com} = h_4 - h_3 = 257 \text{ kJ/kg}$$

(b) The COP is COP = q_e/w_c = 4.13.

(c) The pressure ratio amounts to P_4/P_3 = 5.38.

(d) The volumetric capacity, v_{vol}, is a measure for the amount of cooling capacity per unit of volume flow through the compression, v_3 is the specific volume of the suction vapor.

$$v_{vol} = q_e/v_3 = (1060 \text{ kJ/kg})/(0.42 \text{m}^3/\text{kg}) = 2524 \text{ kJ/m}^3$$

(e) The temperature at the compressor outlet is 118°C as described above.

Remark: The energy balance for the entire cycle is applied to check for consistency.

$$q_e + w_{com} - q_c = 0 = 1060 \text{ kJ/kg} + 257 \text{ kJ/kg} - 1317 \text{ kJ/kg}$$

In the following, various options are listed to increase the efficiency by internal heat transfer. For example, in the heat pump, the liquid flowing from B to A can be subcooled by the gas leaving A, as indicated in Figure 3.8. The cooling capacity delivered in A is increased. The compressor suction gas is superheated, requiring usually higher work input. Therefore, for each working fluid, the net gain has to be carefully investigated.The net gain is negative for ammonia and water and positive for fluids like R114 and R134a.

In case of the heat engine, the liquid pumped from A to B can be preheated by the condensation of partially expanded steam bled from the turbine (feed water preheating). The reverse process may be used with heat pumps when the compressor has two stages (Figure 3.9).

Also, the superheat of the compressed working fluid after the first stage is reduced by evaporating some of the liquid working fluid on the same pressure level.

3.3 CONDENSATION AND EVAPORATION OF WORKING FLUID MIXTURES

When mixtures are used in Rankine Cycles, similar phase changes occur between a condensed phase and a vapor phase as in pure components (Figure 3.1). Figure 3.10 displays examples for the working fluid water. However, the process termed "evaporation" is now called "desorption", and the process of "condensation" is termed "absorption". For the sake of simplicity, it is assumed that only one component changes its phase. The component remaining liquid or solid is called the absorbent. There are many reactions analogous to those mentioned in Figure 3.10 involving gases like NH_3, CO_2, hydrogen, halogenated hydrocarbons, etc. (Figure 3.11).

In the literature, a variety of expressions (Figures 3.10 and 3.11) show liquid and solid absorbents. Furthermore, Figure 3.10 shows reactions in which the vapor pressure line of the mixture or compound (left-hand side) is either a set of vapor pressure lines (solution filed) or a single vapor pressure line. Some of the solid-vapor reactions can be classified as adsorption or absorption. However, neither the type of reaction nor the type of solution field is relevant to achieve the heat pump effect. Therefore, a classification in terms of absorption, adsorption, vapor-liquid, solid-vapor, or chemical heat pump is, in the best case, a mere indication of the required engineering design. It may also provide a hint about the nature of the kinetic processes.

Several important differences can be found when comparing the evaporation of a pure component with the desorption out of an absorbent:

A. The presence of the absorber causes a reduction of vapor pressure as a function of temperature or boiling point elevation (or elevation of the temperature for sublimation). When desorption or absorption occurs under constant pressure from a solution, the "boiling temperature" or "condensing temperature" of the mixture is higher than that of the pure fluid. During isothermal processes, the saturation pressure changes according to the composition of the mixture.

B. The phase changes vapor to liquid or vapor to solid can occur in temperature and pressure ranges which are far above the critical temperature or critical pressure of the pure working fluid.

The properties mentioned above under A are important for new heat pump processes and potentially for new power generation cycles using mixtures.

Figures 2.7 and 2.15 (Chapter Two) exhibit the vapor pressure curves of the mixtures NH_3/H_2O and $H_2O/LiBr$, while Figure 11.5 (Chapter Eleven) shows the vapor pressure lines for H_2O in equilibrium with a zeolite. For the stoichiometric reactions of Figures 3.10 and 3.11, only one vapor pressure curve exists, according to Gibb's phase rule. (See Figure 11.4, Chapter Eleven, for $Ca(OH)_2$).

Multicomponent systems in which all components evaporate are also of technical importance. Examples for these mixtures were given in Chapter Two under the term zeotropic mixtures. During the phase change process, the composition of the vapor phase is different from the liquid phase. For constant pressure processes, the phase changes occur in temperature intervals (gliding temperature intervals) which depend on composition. The gliding temperature intervals that are found during the phase change of zeotropic mixtures in evaporators and

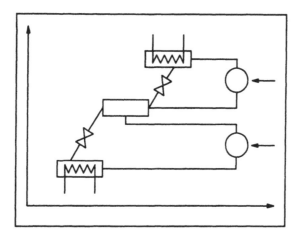

Figure 3.9 Cooling of a liquid stream and of superheated vapor by evaporation.

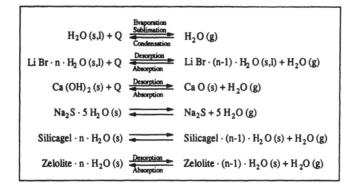

Figure 3.10 Examples for "evaporation" and "condensation" of water, (g, vapor; f, liquid; s, solid).

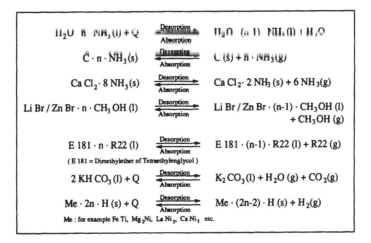

Figure 3.11 Examples for "evaporation" and "condensation" of water, (g, vapor; f, liquid; s, solid).

condensers can be used for increasing the coefficient of performance. This is demonstrated later in Example 3.3. Furthermore, the capacity (i.e., the pressure) can be controlled by adjusting the composition of the circulating fluid. For a constant volume flow rate compressor, the mass flow rate depends on the vapor density of the refrigerant. The vapor density is, in turn, a strong function of the pressure, which depends on the fluid composition. Figure 2.11 (Chapter Two) shows an ln(P),h diagram for a zeotropic fluid mixture.

Example 3.3: Vapor-Compression Heat Pump with R22/R114

A vapor-compression heat pump working with a mixture of R22,R142b (50 mass %) is evaluated. The end temperatures of the evaporation and the condensation process are −10°C and +40°C, respectively. An idealized Rankine Cycle with isentropic compression, no pressure drop in the heat exchangers, and isenthalpic expansion process is assumed.

(a) What is the amount of the evaporator heat, the condenser heat, and the work input to the compressor per kilogram of refrigerant?

(b) What is the COP based on the results of (a)?

(c) How large is the pressure ratio?

(d) How high is the temperature of the vapor at the compressor outlet?

(e) How large are the gliding temperature ranges in the evaporator and the condenser?

(f) Compare the results with a refrigeration system operating with pure R22 achieving the same lowest and highest temperature in the evaporator and condenser.

Solution: First, the state points of the fluid mixture must be found on the ln(P),h-chart, as indicated on Figure 3.12. The actual data are read from Figure 2.12. The end point of the condensation is point 1. It is located on the saturated liquid line at the intersection with the isotherm of 40°C. Point 3 is the evaporator exit, located on the saturated vapor line at the intersection with the isotherm of −10°C. The evaporator inlet is point 2, obtained through isenthalpic expansion from point 1 to the low pressure isobar. Point 4 is the compressor outlet, located on the same isentrope as the evaporator outlet, point 3. Reading the properties at the state points from Figure 2.11, one obtains Table 3.3:

Point 2′ represents the saturated vapor in the condenser. Note: Because of the nonazeotropic character of the mixture, the saturation temperature of the evaporating working fluid changes from −16°C to −10°C, and for the condenser from 53°C to 40°C.

(a) The evaporator and condenser heat and the compressor work are found to be

$$q_e = h_3 - h_2 = 158 \text{ kJ/kg}$$

$$q_c = h_4 - h_1 = 201 \text{ kJ/kg}$$

$$w_{com} = h_4 - h_3 = 43 \text{ kJ/kg}$$

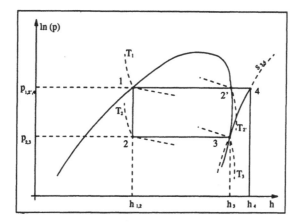

Figure 3.12 Schematic pressure-enthalpy diagram for a zeotropic mixture with superimposed vapor compression cycle.

Table 3.3 Thermodynamic Properties of the State Points

	P(kPa)	T(C)	x	h(kJ/kg)
1	1120	40	0.5	98
2	160	−16	0.5	98
3	160	−10	0.5	256
4	1120	66	0.5	299
4'	1120	53	0.5	280

(b) The COP for cooling is then COP = q_e/w_{com} = 3.67.

(c) From Table 3.3, the compression ratio is found to be P_2/P_1 = 7.0.

(d) The temperature at the compressor outlet is 66°C.

(e) The temperature glide in the condenser is 13 K, including the superheat, and for the evaporator 6 K. The interval in the evaporator is considerably smaller than in the condenser because the flashing in the expansion valve already produces a fraction of refrigerant vapor.

The energy balance yields:

$$q_e + w_{com} - q_c = 0 = 158 \text{ kJ/kg} + 43 \text{ kJ/kg} - 201 \text{ kJ/kg}$$

Remark: The gliding temperature interval in the evaporator can be extended to lower temperatures when the liquid leaving the condenser is first brought into heat exchange with the evaporating fluid in the evaporator before it enters the expansion valve.

(f) The performance of the R22 system is calculated as follows:

$$q_e = h_3 - h_2 = 130 \text{ kJ/kg}$$

$$q_c = h_4 - h_3 = 177 \text{ kJ/kg}$$

$$w_{com} = h_4 - h_3 = 47 \text{ kJ/kg}$$

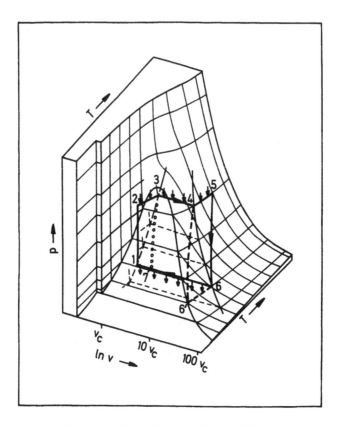

Figure 3.13   The Rankine Sorption Cycle.

The COP for cooling is then COP = q_e/w_{com} = 2.76, and the compression ratio is found to be P_2/P_1 = 8.57. The temperature at the compressor outlet is 81°C. A comparison of this result with that of the mixture of R22/R142b demonstrates the potential for energy savings that comes with the use of the temperature glide. In this example, it was implicitly assumed that the temperature glide is linear. This is not necessarily the case. Also, to take advantage of the temperature glide, counterflow heat exchangers have to be used.

3.4 RANKINE SORPTION CYCLES WITH INPUT AND OUTPUT OF WORK

Figure 3.13 shows the Rankine Sorption Cycle, including the phase changes of desorption and absorption. The new surface of the coexisting thermodynamic states of the condensed and gaseous phases is moved at given pressures to higher temperatures or at given temperatures to lower pressures in comparison to the pure working fluid. The sequence of states within a Rankine Sorption Cycle is the following: 1–2 pressurization, 2–3 heating (temperature increase) of the solution, 3–4 desorption, 4–5 superheating, 5–6 isentropic expansion, 6–1 absorption. Since the vapor at state 4 is superheated with respect to its own condensate, the expansion may be conducted using the process indicated by line 4–6′ without the presence of condensate in the turbine. In order to achieve a reversible cycle, the low pressure steam has to be heated from points 6′ and 6 at constant pressure. Figure 3.14 (a-g) shows the work generating Rankine Sorption Cycle in various projections. Compared to the simple Rankine Cycle for given pressures P_1 and P_2, the temperatures for the heat supplied and rejected are shifted to higher values. For given temperatures T_1 and T_2, the pressures P_1 and P_2 are reduced.

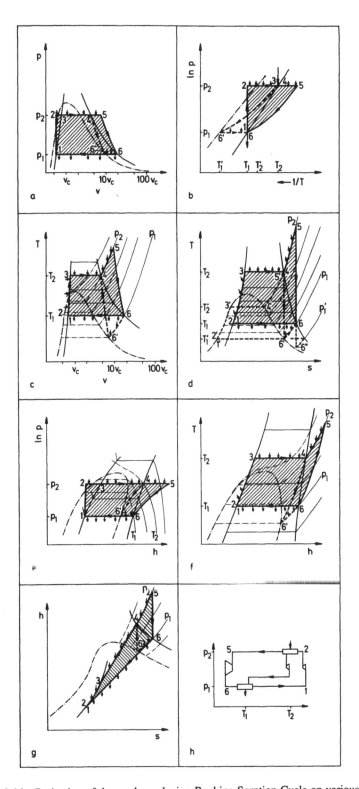

Figure 3.14 Projection of the work-producing Rankine Sorption Cycle on various planes.

A comparison of Figure 3.2h with 3.14h shows that an additional connection is required between the desorber at high pressure and temperature and the absorber at low pressure and temperature. This connection allows for the solution not evaporated in the desorber to return to the absorber. Clearly, this is only possible when the absorbent fluid is a liquid which can be pumped. In the case of solid absorbents, the desorber and absorber alternate their functions as required. In this case a quasi-continuous operation can be achieved by operating two absorber/desorber sets in alternating cycles.

The version of the Rankine Sorption Cycle according to Figure 3.15 (a-h) requiring input of work is called the "vapor-compression heat pump with solution circuit." The potential of such sorption-compression heat pumps for which the pressure level can be reduced for a given temperature range is only beginning to be explored.

Figures 3.13 to 3.15 are valid only for such cycles, where the change of composition in absorption and desorption is small. In the case of a significant composition difference, the shift of the vapor pressure curve 4–6 as a function of composition must be taken into consideration. In diagrams showing desorption and absorption at constant pressure, the lines 3–4 and 6–1 extend over a finite temperature interval (Figure 3.16a,b).

3.5 VAPOR COMPRESSION HEAT PUMP OR HEAT ENGINE WITH SOLUTION CIRCUIT

In Figure 3.17, the scheme of a vapor compression heat pump with a solution circuit is superimposed on the vapor pressure curves of a mixture. Components A and B contain a mixture of absorbent and working fluid. A portion of the working fluid circulates through compressor K.

This type of cycle always requires two connections between A and B. One must have an expansion device and one must have a pump for circulating the liquid phase. For good performance, the two liquid streams circulating between A and B must exchange heat with each other. Without this internal heat exchange, an undesirable heat flow would exist between A and B.

Compared to a conventional vapor-compression heat pump operating with the same working fluid, the cycle shown in Figure 3.17 exhibits the following differences:

A. The pressures P_1 and P_0 are lower for the same temperatures T_1 and T_0.

B. The compressor suction gas is superheated. The absorber A receives superheated working fluid vapor which may be in thermodynamic equilibrium with the absorbent

C. Both components A and B offer the opportunity of heat transfer at gliding temperatures, allowing for a match of the temperature profiles of the working fluids with that of the heat transfer fluid (for example, water or air) across the heat exchangers. This may lead to a significant improvement of the coefficient of performance (Lorenz Cycle), as is illustrated in the following example.

   The return water of a district heating system has a temperature of 50°C. Half of the stream is cooled to 20°C, while the other half is heated to 80°C. The Carnot efficiency for a heat pump that uses a pure working fluid is about 5.9. This value is based on the efficiency for heating. Using gliding temperatures, the effective temperature difference between the high temperature heat sink and the low temperature heat source is reduced from 60 K to 30 K. Consequently, the coefficient of performance amounts to 10.3, approximately twice as large as for the pure fluid system.

D. By changing the composition of the solution circulating between A and B at constant temperatures, T_0 and T_1, the saturation pressures can be altered. The compressor K is supplied with lower or higher density vapor resulting in a capacity change at a constant

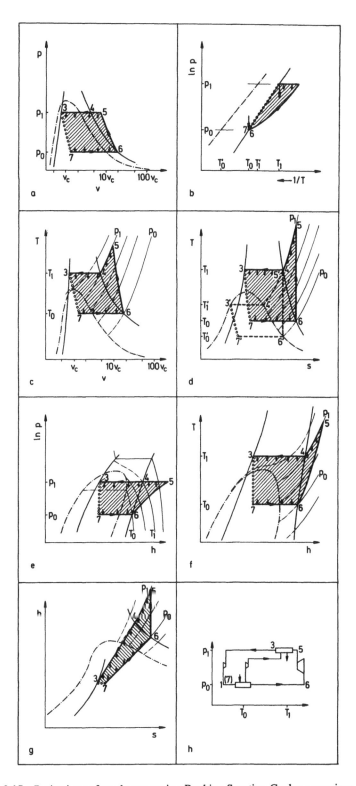

Figure 3.15 Projections of work-consuming Rankine Sorption Cycles on various planes.

rotary velocity. Figure 3.18 shows an example for additional components which may be required for changing the composition. G_1 is the main receiver for solution that is low in refrigerant, and G_2 collects absorbent only. In order to avoid undesired storage of refrigerant in G_2, the solution that is low in refrigerant (but at a high temperature

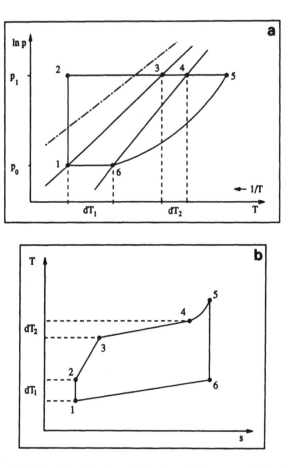

Figure 3.16 Projection of the Rankine Sorption Cycle for a finite composition difference: (a) pressure-temperature diagram; (b) temperature-entropy diagram.

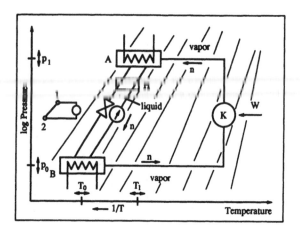

Figure 3.17 Vapor-compression heat pump with solution circuit in $\ln(P),(-1)/T$ diagram.

when coming from the absorber) is collected and cooled further by evaporating refrigerant drawn by the compressor through a check valve. G_2 may be provided with a low capacity heating element to further desorb refrigerant. The composition change for capacity control seems to be simpler for this cycle than for conventional ones with zeotropic mixtures.

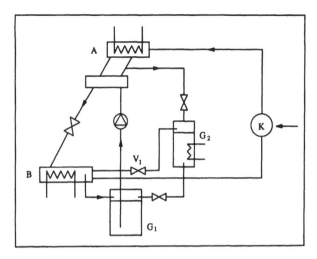

Figure 3.18 Schematic diagram for capacity control of a vapor-compression heat pump with solution circuit.

When the directions of all fluid and energy streams are reversed, as shown in Figure 3.17, the cycle represents a heat engine with solution circuit. The fraction of the temperature glide is preserved.

In the past, Organic Rankine Cycles have been used to match the temperature glide of the heat source fluid. When compared to water, these fluids allow not only smaller volume flow rates, but also a well-tailored temperature distribution in the boiler matching that of the heat source fluid. This leads to reduced irreversibilities, Figure 3.19. A match is not possible for condensation (Figure 3.19).

The Rankine Sorption Cycle exhibits gliding temperatures for absorption and desorption (Figure 3.20). In moving from state 2 to state 3, only the difference of the sensible heat between the two solution streams must be provided, since the solution with the higher working fluid concentration is preheated by means of the counter flow heat exchanger H (Figure 3.17).

During the desorption process, working fluid vapor is generated. This vapor is superheated with respect to its own condensate. Therefore, no superheater may be required.

The pressure of the mixture is always lower than for the pure working fluid. The degree of vapor pressure reduction depends on the concentration. The cycle can be operated at temperature levels higher than the critical temperature of the working fluid. This is frequently done using NH_3, for instance. An example for the use of ammonia/water sorption processes for the power generation providing the match of temperature glides is referred to as "Kalina Cycle".

In order to achieve capacity control, similar means can be employed, as discussed in Figure 3.18. In Chapter Nine, cycles are discussed which allow the conversion of waste heat below 100°C by using the well-proven working fluid H_2O, but the turbine is supplied with steam at pressure levels higher than 1 bar.

Example 3.4: Vapor-Compression Heat Pump with Solution Circuit

A vapor-compression heat pump operating with a solution circuit is evaluated. The working fluid mixture is $NH3/H_2O$ with 40 and 50 weight % of NH_3 in the strong and weak absorbent. The end temperatures of the absorption and of the desorption processes are −10°C and +40°C. It is assumed that the compression in the liquid and vapor phase is isentropic. There is no pressure drop in the heat exchangers and connecting piping, and the expansion process is isenthalpic.

Answer the following questions based on 1 kg of circulating refrigerant vapor:

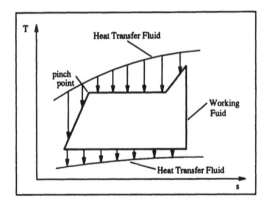

Figure 3.19   Heat transfer for pure working fluids.

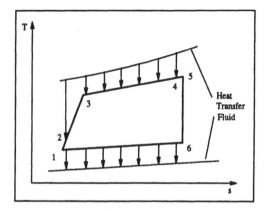

Figure 3.20   Heat transfer, when mixture is used.

(a) What is the flow rate of solution rich in ammonia?

(b) What is the pump work input and the temperature of the liquid leaving the pump?

(c) How much heat is exchanged in the solution heat exchanger?

(d) What is the compressor work input and the temperature of the discharge vapor?

(e) What is the amount of heat exchanged in the desorber and absorber?

(f) What is the COP of the system?

(g) How large is the pressure ratio?

(h) How large is the volumetric capacity?

(i) What is the maximum possible value of the gliding temperature ranges in desorber and absorber?

Solution: (a) First the state points of the fluid streams must be found on the h,x-diagram for ammonia and water, Figure 2.2, (Chapter Two). (Note: The diagram in Figure 2.2 shows the pressure in bar and the temperature in K). A schematic diagram of the h,x diagram is shown in Figure 3.21, and a schematic diagram of the cycle is shown in Figure 3.22. The end point of the

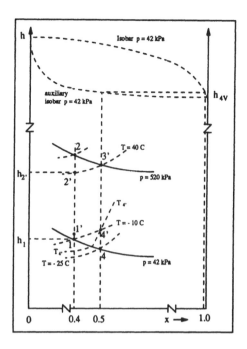

Figure 3.21   Schematic enthalpy-concentration diagram.

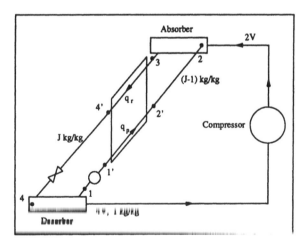

Figure 3.22   Schematic diagram of vapor-compression cycle with solution circuit. The numbers of the state points refer to Figure 3.21.

desorption process, point 1, is located where the isotherm of –10°C intersects with the line of constant concentration of 0.4. The pressure at this point is 43 kPa. Point 3, the end point of absorption, is found at the intersection of the 40°C isotherm and the isostere of 0.5. The pressure at this point is 547 kPa. Point 2, the starting point of the absorption, if the liquid entering the absorber were saturated, is found where the same isobar of 547 kPa, as determined from point 3, intersects with the isostere of 0.4. This state point is usually not obtained in a realistic process. Point 4 is found at the intersection of the low pressure isobar through point 1 of 43 kPa with the isostere of 0.5. Point 4 represents the beginning of the desorption process, when there would be no flashing in the expansion

Table 3.4   Thermodynamic Properties
of the State Points

		P(kPa)	T(C)	X(kg/kg)	h(kJ/kg)
1	Desorb. outlet	43	-10	0.40	-289
1'	Pump outlet	547	-10	0.40	-288
2	Sat.liq.in abs.	547	58	0.40	30
2'	Abs. inlet	547	40	0.40	-59
2_v	Comp. outlet	547	191	1.00	1672
3	Absorber outlet	547	40	0.50	-61
4	Sat. liq. in des.	43	-24	0.50	-359
4_v	Des. vapor	43	-10	1.00	1242
4'	Exp. valve in	547	-10	0.50	-300
4"	SHX out	547	0	0.50	-252

Note: The diagram in Figure 2.2 lists the pressure in bar and the temperatures in K.

valve. The temperature at which the actual desorption process begins is $T_{4'}$.
All property data for the state points above and the ones found later are
listed in Table 3.4. The calculations are performed for 1 kg of vapor
flowing through the compressor. The concentration of the vapor leaving
the desorber, x_v has to be determined first. To find x_v from the diagram, it
is assumed that the vapor is either in equilibrium with the liquid at point
1 or at point 4. The latter assumption is made here. The vapor in equilibrium with the liquid phase at point 4 contains less water, and therefore the
assumption that the vapor is pure ammonia, which has to be made later, is
more accurate. x_v is found at the intersection of the line of constant
concentration $x_4 = 0.50$ with the auxiliary isobar of 43 kPa in the vapor
phase. From there, one follows a line of constant enthalpy to the intersection with the dew point line of 43 kPa. In this case, the vapor concentration
is found to be pure ammonia within the accuracy of reading the diagram,
$x_{4v} = 1.0$. The enthalpy of the vapor is $h_{4v} = 1242$ kJ/kg.

The flow rate of the solution rich in ammonia, f, is found according
to the following equation:

$$f = (x_v - x_1) / (x_3 - x_1)$$

This results from the application of the mass balances for the ammonia
and total mass for the desorber, Figure 3.23

$$(f-1) x_1 + 1 x_v = f x_3$$

For f, the following value is found:

$$f = (1.0 - 0.4)/(0.5 - 0.4) = 6.0$$

The equation for f is only dependent on concentrations.

(b) Next, the enthalpy change of the liquid of point 1 is calculated as it passes
through the solution pump. The work input to the pump amounts to:

$$w_p = (f-1)(P_2 - P_1) v_s$$

$$w_p = (6.0 - 1)(547 \text{ kPa} - 43 \text{ kPa}) 1/840 \text{ m}^3 \text{ kg} = 3.0 \text{ kJ/kg}$$

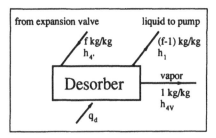

Figure 3.23 Schematic diagram of a desorber.

where $f - 1$ is the flow rate through the pump, $P_2 - P_1$ is the pressure difference the pump has to overcome, and v_s is the specific volume of the fluid being pumped. The specific volume is the inverse of the density. Here a value of 840 kg/m³ is assumed since this is the concentration averaged value based on the density of 1.0 kg/m³ for water, 600 kg/m³ for ammonia. This value is only an estimate. However, since the pump work is very small compared to any other amount of energy exchanged in the system, the estimate is acceptable. The enthalpy of the liquid entering the solution heat exchanger amounts to:

$$h_{1'} = h_1 + w_p /(f - 1)$$

The enthalpy $h_{1'}$ is found to be

$$h_{1'} = -289 \text{ kJ/kg} + 3.0 \text{ kJ/kg}/5.0 = -288 \text{ kJ/kg}$$

The temperature of the liquid after the pump can, in principle, be read off Figure 2.2. However, the change in enthalpy is smaller than can be resolved in the diagram. The temperature change is negligible.

(c) With this information, the performance of the solution heat exchanger can be evaluated. First, the enthalpies $h_{2'}$ and $h_{4'}$ are found. They are located on the same isosteres as h_2 and h_4, but at the intersection with isotherms of 40°C (for $h_{2'}$) and 10°C for $h_{4'}$. It is here considered that the stream to be heated (or cooled) cannot become warmer (or colder) than the stream that supplies (absorbs) heat. Thus $h_{2'} = -59$kJ/kg and $h_{4'} = -300$kJ/kg are found. The heat capacity of both streams, q_r and q_p, in the solution heat exchanger is calculated:

$$q_r = f(h_3 - h_4) = 6.0((-61 \text{ kJ/kg}) - (-300 \text{ kJ/kg})) = 1434 \text{ kJ/kg}$$

and

$$q_p = (f - 1)(h_{2'} - h_{1'}) = (6.0 - 1.0)((-59 \text{ kJ/kg}) - (-288 \text{ kJ/kg})) = 1145 \text{ kJ/kg}$$

The amount of heat exchanged has to be the same for both heat exchanger paths and is equal to the smaller one resulting from q_r and q_p. Here, q_p is the smaller one. Therefore, the enthalpy $h_{4'}$ has to be recalculated so that q_r and q_p are equal. From the requirement that q_p and q_r are equal, it follows:

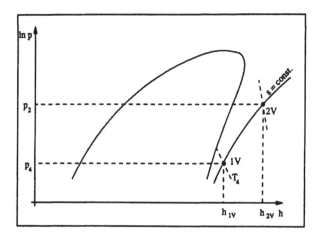

Figure 3.24   Schematic pressure-enthalpy diagram to be used in conjunction with Figure 3.21.

$$-h_{4^-} = (f-1)\left(h_{2'} - h_{1'}\right)/f - h_3$$

$$h_{4^-} = -5.0\left((-59 \text{ kJ/kg}) - (-288 \text{ kJk/g})\right)/6.0 + (-61 \text{ kJ/kg}) = -252 \text{ kJ/kg}$$

This result illustrates why the liquid stream from 3 to 4 is not cooled to the degree that would be desirable, because the stream from 1 to 2 does not provide sufficient heat capacity. This is to be expected, because the flow rate is smaller than that of the stream from 3 to 4. The temperature for 4″ can be read of Figure 2.2 and is found to be about 0°C.

(d) The work input to the compressor is found with the help of the ln(P),h chart for pure ammonia. A schematic diagram is shown in Figure 3.24. The actual data are obtained from Figure 2.8, Chapter Two. Care has to be taken that the vapor enthalpies are correctly based on the enthalpies on the h,x-chart. The ln(P),h-chart may have a different reference point for the enthalpy. This is the case here. To be consistent, only enthalpy differences are transferred from one chart to the other. With the known temperature and pressure of the suction vapor, −25°C and 43 kPa, its enthalpy can be found on the ln(P),h diagram in the superheated region to be 480 kJ/kg. Following the isentrope in the isobar P_2, 547 kPa, the enthalpy for the compressor outlet and its temperature can be read as 910 kJ/kg and 191°C. The difference in these two values is the compressor work input for isentropic compression, w_{com} = 430 kJ/kg. The value of the enthalpy of the vapor entering the absorber can now be calculated as 1242 kJ/kg + 430 kJ/kg = 1672 kJ/kg.

(e) The heat exchanged in the desorber, q_d, and absorber, q_a, are calculated now. An enthalpy balance for the desorber yields, Figure 3.23:

$$q_d = h_{4v} - h_1 + f\left(h_1 - h_{4^-}\right)$$

$$q_d = 1242 \text{ kJ/kg} - (-289 \text{ kJ/kg}) + 6.0\left((-289 \text{ kJ/kg}) - (-252 \text{ kJ/kg})\right) = 1309 \text{ kJ/kg}$$

The enthalpy balance for the absorber yields:

$$q_a = h_{2v} - h_{2'} + f\left(h_{2'} - h_3\right)$$

$$q_a = 1672 \text{ kJ/kg} - \left(-59 \text{ kJ/kg}\right) + 6.0\left(\left(-59 \text{ kJ/kg}\right) - \left(-61 \text{ kJ/kg}\right)\right) = 1743 \text{ kJ/kg}$$

To check for consistency, the energy balance is applied:

$$q_d - q_a + w_{com} + w_p = 0$$

$$1309 \text{ kJ/kg} - 1743 \text{ kJ/kg} + 430 \text{ kJ/kg} + 3 \text{ kJ/kg} = -1 \text{ kJ/kg}$$

The discrepancy of −1 kJ/kg results from round-off errors and, compared to any amount of energy exchanged in the system, it is negligible.

(f) The COP of the system amounts to:

$$COP = q_d / \left(w_{com} + w_p\right) = \left(1309 \text{ kJ/kg}\right) / \left(430 \text{ kJ/kg} + 3 \text{ kJ/kg}\right) = 2.77$$

(g) The pressure ratio amounts to 547 kPa/43 kPa = 12.7

(h) The volumetric capacity amounts to (1309 kJ/kg)/(0.36 kg/m³) = 3636 kJ/m³. The value for the specific volume of the vapor is obtained from the ln(P),h diagram, Figure 2.8, Chapter Two.

(i) From Table 3.4, the maximum possible gliding temperature interval is 24 K for the desorber and 18 K for the absorber. The maximum temperature glide is only obtained when the saturation temperatures at points 2 and 4 are actually achieved. This is not the case. The fluid leaving thesolution heat exchanger entering the absorber has, at best, reached the temperature of the fluid exiting the absorber. Thus the absorption process starts with the temperature T_3. But the heat developed during the absorption process raises this temperature to some intermediate level between T_2 and T_3. The actual temperature depends on the heat exchanger design and the operating conditions. Since the fluid enters the absorber through the expansion device, a fraction is already evaporated. Thus, the lowest temperature T_4 cannot be attained. The actual temperature is determined by the two-phase isotherm through point 4' for the pressure of P = 43 kPa. The location of this isotherm is indicated in Figure 3.21.

Remarks:

(1) The calculation procedure exemplified above demonstrates clearly the sequence (questions a through f) one should adhere to. This is the fastest method to calculate all relevant parameters of a solution circuit.

(2) The assumption that the vapor leaving the desorber is in equilibrium with the liquid at point 4 is only an approximation. Since the expansion process occurs at constant enthalpy point 4' is located in the two-phase region. Some fraction of vapor is already present, and the vapor concentration cannot be the same as the equilibrium vapor for a liquid at point 4. However, this discrepancy is negligible in the case presented here. If the vapor concentration changes significantly, one has to redo the evaluation with the new concentration or use a more appropriate assumption.

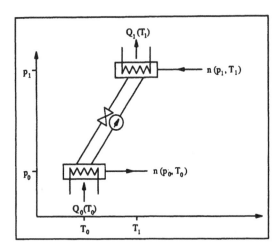

Figure 3.25 Pair of exchange units.

3.6 TRANSFORMATION OF AVAILABILITY BY PAIRS OF EXCHANGE UNITS

A compressor converts work into the availability of a high pressure gas stream. An expansion turbine converts the differences of the availability between a high and a low pressure gas stream into work. This process is independent of how the availability increase of the high pressure gas stream is achieved.

Two exchange units (heat exchanger with phase change of the working fluid) which operate on different pressure levels and which are connected by piping for the transfer of condensed working fluid (with or without absorbent), Figure 3.25, may also be used as a conversion device for availability.

An exchange unit may be defined as an apparatus in which working fluid undergoes a phase change combined with the exchange of heat. The pair of exchange units shown in Figure 3.25 operates with a fluid stream in the direction of the arrows, entering at high pressure and leaving at low pressure. This pair of exchange units transfers the difference in availability of the fluid stream entering at high pressure and leaving at low pressure into a difference of availability of a heat flow entering at a low temperature and leaving at high temperature.

This change is represented by:

$$\text{Availability } (n(p_1, T_1), n(p_0, T_0)) \leftrightarrow \text{Availability } (Q_1(T_1), Q_0(T_0)) \tag{3.1}$$

(n = working fluid flow rate)

When the flow directions are reversed, the difference of availability of a heat flow is converted into the difference of availability of a working fluid stream.

These properties of counterclockwise and clockwise operating exchange units are used to build absorption heat pumps and heat transformers. For example, it has been proposed to use hydrogen absorbing metals to separate hydrogen from a gas mixture and to produce pure hydrogen on a high pressure level by applying high temperature heat. This process employs the principle depicted in Figure 3.25.

When in the device in Figure 3.25 only pure working fluid and no absorbent is circulating, then the two exchange units represent the condenser and evaporator of a vapor compression heat pump or the condenser and boiler of a heat engine.

3.7 COP OF COMPRESSOR HEAT PUMPS BASED ON THE SECOND LAW

To calculate the COPs of compressor heat pumps or refrigerators, complete enthalpy-pressure or enthalpy-entropy or temperature-entropy diagrams must be available. By applying the First

Law of Thermodynamics and using the enthalpy charts, the COP can be determined (First Law method). However, the COP of a heat pump must and will obey the Second Law of Thermodynamics. By using the experimentally determined property charts of real fluids, the requirements of the Second Law of Thermodynamics are taken into account inherently, in an indirect way. The term "First Law efficiency" for a COP is thus misleading.

The statements of the Second Law of Thermodynamics are independent of fluid properties. It is unsatisfactory that the Second Law must be recalculated for each case considered, requiring an extensive amount of experimental data with great accuracy. Therefore, in this chapter, a different approach is presented: the Second Law of Thermodynamics is taken into account from the very beginning. Introducing the requirements of the Second Law in a direct way, the required knowledge about fluid properties is reduced to a few nondimensional parameters. These parameters can be used to characterize and evaluate fluids.

The entropy is a state function. Thus, entropy balances can be used. The method that is presented here should not be confused with an availability or exergy analysis, for which the term "Second Law analysis" is commonly used. On the contrary, it will be shown that this analysis does not have the shortcomings of the exergy analysis. It is not dependent on a poorly defined environmental temperature.

3.7.1 COP η for Refrigeration Using the Second Law

Figure 3.26 shows schematically the energy flow in a vapor compression refrigerator or heat pump. The lower case letters are used for the external temperatures, i.e., those of the heat transfer fluids, and capital letters are used for the internal temperatures, i.e., the refrigerant temperatures. In the case of a working fluid mixture, the internal temperatures have a certain temperature glide. In Figures 3.27 and 3.28, the compression process is displayed for fluids with different properties at the dew line. For R22, for example, $ds/dT < 0$ or for R114, $ds/dT > 0$.

The First and Second Laws of Thermodynamics can be written as follows:
First Law:

$$w + q_1 = q_2 \tag{3.2}$$

Second Law:

$$\frac{q_2}{t_2} - \frac{q_1}{t_1} = \sum_{i=1}^{n} \delta s_i \tag{3.3}$$

where $\Sigma \delta s_i$ (the irreversible increase in entropy), q_1, q_2, and w are larger than zero. All quantities are actually time derivatives (fluxes) and refer to one unit of mass of circulating working fluid. The right-hand side of Equation 3.3 is the sum over all the entropy-creating steps of the cycle, including the heat transfer to the external fluids. All δs_i contributions are positive. The terms q_2/t_2 and q_1/t_1 are the entropy changes per unit time for the heat transfer fluids and are determined as follows (see Figure 3.26):

$$q_i = \int_{t_i^1}^{t_i^2} c_i dt$$

$$\frac{1}{t_i} = \frac{1}{q_i} \int_{t_i^1}^{t_i^2} \frac{c_i dt}{t} \tag{3.4}$$

where c_i is the heat capacity of the transfer fluid.

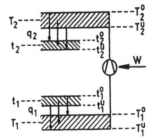

Figure 3.26 Internal temperature T and external temperature t and heat flow for a compressor heat pump. T_i and t_i are entropic averages.

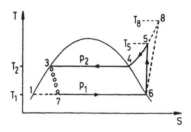

Figure 3.27 Temperature-entropy diagram for a fluid with $ds/dt < 0$, i.e., $R\beta/c_p > 1$ at the dew line.

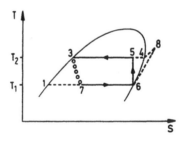

Figure 3.28 Temperature-entropy diagram for a fluid with $ds/dt > 0$, i.e., $R\beta/c_p < 1$ at the dew line.

Temperatures t_1 and t_2 are precisely determined averages, called entropic averages. For most cases at and above room temperature, the arithmetic average temperature is a good approximation:

$$t_i \approx \left(t_i^1 + t_i^2\right)/2 \qquad (3.5)$$

The following irreversible process steps can be identified:

δs_1 = throttling 3–7;

δs_2 = desuperheating between 5 and 4, i.e., flow of heat to the temperature T_2 after isentropic compression;

δs_3 = compression between 6 and 8 desuperheating between 8 and 5, i.e., flow of that part of the heat created by irreversible compression to the temperature level T_2;

δs_4 = heat transfer at the condenser, i.e., between T_2 and t_2;

δs_5 = heat transfer at the evaporator, i.e., between t_1 and T_1;

δs_6 = pressure drops by, e.g., flow in pipes;

δs_7 = subcooling of liquid fluid and superheating of gaseous fluid;

δs_8 = open-ended listing.

The individual contributions, δs_i, are not independent of each other. For example, the irreversibilities of the compressor are increasing q_2 and thus δs_4 but not δs_5. A change in δs_1 (for example, by using an expansion turbine or a different fluid) influences the value of δs_5.

Now the COP, η, and quality factors for refrigeration are determined. A conversion to COPs for heat pumping is made in the next section.

Eliminating q_2 from Equations 3.2 and 3.3 yields the following equations for the work and the COP for refrigeration:

$$w = q_1 \frac{t_2 - t_1}{t_1} + t_2 \sum_1^n \delta s_i \tag{3.6}$$

$$\eta = \frac{\eta_r(t_1, t_2)}{1 + \frac{t_1 t_2}{t_2 - t_1} \frac{\Sigma \delta s_i}{q_1}} \tag{3.7}$$

where $\eta(t_1,t_2) = t_1/(t_2 - t_1)$ is the reversible or Carnot efficiency for refrigeration with respect to the external temperatures. Reversible vapor compression processes ($\Sigma \delta s_i = 0$) operate according to the Carnot efficiency:

$$\eta(\text{rev}) = \eta_r(t_1, t_2) \tag{3.8}$$

with temperatures t_2 and t_1 defined in Equation 3.4.

The quality factor $g(t_1,t_2)$ is defined as the ratio of η_r for the real process divided by η for a reversible process, i.e., a process with infinitely large heat exchanger surfaces, an isentropic compressor, working with a hypothetical fluid with negligible superheating and throttling losses.

$$g(t_1, t_2) = \frac{q_1}{q_1(\text{max})} = \frac{\eta}{\eta(\text{rev})} = \frac{1}{1 + \frac{t_1 t_2}{t_2 - t_1} \frac{\Sigma \delta s_i}{q_1}} \tag{3.9}$$

The irreversibilities $\Sigma \delta s_i$ are weighted by process temperatures t_1 and t_2, in contrast to similar quality factors derived from the exergy concept into which an ambient temperature, t_u, enters. If Equation 3.6 is written as:

$$w = w_0(t_1, t_2) + \Sigma w_i \tag{3.10}$$

one finds for the quality factor $g(t_1, t_2)$:

$$g(t_1, t_2) = \frac{w_0(t_1, t_2)}{w} \tag{3.11}$$

According to Equations 3.6 and 3.7, the work required to produce a certain refrigeration capacity q_1 is determined by the minimum value $w_0 = q_1 (t_2 - t_1)/t_1$ and additional work caused by irreversibilities. This extra work is determined by $t_2 \Sigma \delta s_i$ which is not the exergy loss; t_2 can widely differ from the ambient temperature t_u.

The irreversibilities, δs_i, in Equations 3.6 and 3.7 are partly caused by fluid properties, for example δs_1 and δs_2, and partly by design, as the size of the heat exchanger surfaces or properties of the compressor. The irreversibilities dependent on design will be considered first. For simplicity, the entropy increases δs_6 to δs_8 are ignored at present.

For the heat transfer out of the condenser and into the evaporator the following equations hold:

$$\delta s_4 = \frac{q_2}{t_2} - \frac{q_2}{T_2} \tag{3.12}$$

$$\delta s_5 = \frac{q_1}{T_1} - \frac{q_1}{t_1} \tag{3.13}$$

For fluid mixtures, the internal temperatures T_1 and T_2 are exactly defined entropic averages analogous to Equation 3.4. Inserting Equations 3.12 and 3.13 into Equations 3.2 and 3.3 and eliminating q_2, yields:

$$w = q_1 \frac{T_2 - T_1}{T_1} + T_2 \sum_1^3 \delta s_i \tag{3.14}$$

The Carnot factor is now determined by the internal temperatures, T_1 and T_2, and only three entropy sources remain.

The irreversibilities δs_3 caused by the compressor can be calculated in two steps:

$$\delta s_3 (\text{compression}) = s_8 - s_6 \tag{3.15}$$

$$\delta s_3 (\text{desuperheating } 8 - 5) = (h_8 - h_5)/T_2 + s_5 - s_8$$

$$= (1 - g_c) w / T_2 + s_6 - s_8 \tag{3.16}$$

where g_c is the isentropic efficiency or quality factor of the compressor, defined as:

$$g_c = \frac{w(\text{rev})}{w} \tag{3.17}$$

W(rev) is the work required for isentropic compression.
The sum of Equations 3.15 and 3.16 yields:

$$\delta s_3 = \frac{(1 - g_c) w}{T_2} \tag{3.18}$$

This result is plausible: the additional work $(1-g_c)w$ caused by nonisentropic compression finally ends as heat at the temperature level T_2. To calculate the entropy change, the irreversible compression process is replaced by a reversible process and the addition of heat at the temperature level T_2.

With Equation 3.18, the work w and the COP are found:

$$w = \frac{1}{g_c}\left[q_1\frac{T_2 - T_1}{T_1} + T_2(\delta s_1 + \delta s_2)\right]$$

$$= \frac{1}{g_c}\left[w_0(T_1, T_2) + w_1 + w_2\right] \tag{3.19}$$

$$\eta = g_c\frac{\eta_r(T_1, T_2)}{1 + \dfrac{T_1 T_2}{T_2 - T_1}\dfrac{\delta s_1 + \delta s_2}{q_1}} \tag{3.20}$$

where $\eta_r(T_1, T_2) = T_1/(T_2 - T_1)$ is the reversible or Carnot efficiency with respect to the internal temperatures.

The value of η calculated with Equation 3.20 is identical to that resulting from Equation 3.7. Again, a quality factor $g(T_1, T_2)$, is defined as the ratio η for the real process over η for a process with identical internal condenser and evaporator temperatures, with a reversible compressor and working with a hypothetical fluid with negligible superheating and throttling losses, can be defined as:

$$g(t_1, T_2) = \frac{\eta}{\eta(\text{rev})} = \frac{g_c}{\left(1 + \dfrac{T_1 T_2}{T_2 - T_1}\dfrac{\delta s_i + \delta s_2}{q_1}\right)} \tag{3.21}$$

This quality factor is the product of the quality factor of the compressor and that of the fluid g_f with:

$$g_f = \frac{1}{1 + \dfrac{T_1 T_2}{T_2 - T_1}\dfrac{\delta s_1 + \delta s_2}{q_1}} = \frac{w_0(T_1, T_2)}{w_0 + w_1 + w_2} \tag{3.22}$$

The quantities $\delta s_1/q_1$ and $\delta s_2/q_1$ are characteristic for the properties of different fluids; they can either be calculated analytically or taken from property charts according to the following equations:

$$\delta s_2 = \frac{q_{54}}{T_2} - \int_{T_2}^{T_5}\frac{C_p dT}{T}$$

$$= \frac{h_5 - h_4}{T_2} - (s_5 - s_4) \tag{3.23}$$

This term is exactly zero for fluids of the type shown in Figure 3.28.

$$\delta s_1 = \frac{h_3 - h_1}{T_1} - (s_3 - s_1) \tag{3.24}$$

The refrigeration power q_1 is given by:

$$q_1 = h_6 - h_7 = r - \left(h_3 - h_1\right) \tag{3.25}$$

where r is the heat of evaporation per unit of mass at T_1.

Using Equations 3.23 to 3.25 in Equations 3.19 and 3.20, the work, w, and the COP, η, can be calculated as follows:

$$w = \frac{1}{g_c}\left\{\frac{T_2 - T_1}{T_1}q_1 + \frac{T_2}{T_1}\left[h_3 - h_1 - T_1\left(s_3 - s_1\right)\right] + h_5 - h_4 - T_2\left(s_5 - s_4\right)\right\} \tag{3.26}$$

$$\eta = g_c \frac{\dfrac{T_1}{\left(T_2 - T_1\right)}}{1 + \dfrac{T_1 T_2}{T_2 - T_1}\left[\dfrac{h_3 - h_1 - T_1\left(s_3 - s_1\right)}{T_1\left(h_6 - h_3\right)} + \dfrac{h_5 - h_4 - T_2\left(s_5 - s_4\right)}{T_2\left(h_6 - h_3\right)}\right]} \tag{3.27}$$

These equations must yield exactly the same results as the commonly used equations, since no approximation has been made in the derivation. This will be proven at the end of this chapter. Although Equations 3.26 and 3.27 look more complicated at first glance, they have several important advantages which will be discussed in the numerical results section, (3.7.3, below).

3.7.2 COP η_{HP} for Compressor Heat Pumps

The efficiency for heat pumping can be calculated quite analogously to η by eliminating q_1 from Equations 3.2 and 3.3. The entropy changes δs_i are now multiplied by t_1 and divided by the heat of condensation q_2. Yet, since q_2 depends on δs_3 and thus on w, the equations become rather complicated initially. It is easier to calculate η first and from this, η_{HP}.

Using Equation 3.20, the COP for heating η_{HP} can be written as:

$$\eta_{HP} = 1 + g_c \frac{T_1}{T_2 - T_1}\left(1 + \frac{T_1 T_2}{T_2 - T_1}\frac{\delta s_1 + \delta s_2}{q_1}\right)^{-1}$$

$$= \frac{T_2}{T_2 - T_1} - \frac{T_1}{T_2 - T_1}\left[1 - g_c\left(1 + \frac{T_1 T_2}{T_2 - T_1}\frac{\delta s_1 + \delta s_2}{q_1}\right)^{-1}\right] \tag{3.28}$$

with

$$\eta_{HP}(\text{rev}) = \frac{T_2}{T_2 - T_1} \tag{3.29}$$

the quality factor $g_{HP}(T_1, T_2)$ corresponding to Equation 3.21 may now be defined as:

$$g_{HP}\left(t_1, t_2\right) = \frac{q_2}{q_2(\text{max})} = \frac{\eta_{HP}}{\eta_{HP}(\text{rev})} = \frac{1 + \eta}{1 + \eta_r\left(T_1, T_2\right)}$$

$$= 1 - \frac{T_1}{T_2}\left[1 - g\left(T_1, T_2\right)\right] \tag{3.30}$$

A similar relation can be derived from Equation 3.7 for the quality factor $g_{HP}(t_1, t_2)$ which includes the heat transfer:

$$g_{HP}\left(t_1, t_2\right) = \frac{1 + \eta}{1 + \eta_r\left(t_1, t_2\right)} = 1 - \frac{t_1}{t_2}\left[1 - g\left(t_1, t_2\right)\right] \tag{3.31}$$

Table 3.5 Parameters for Three Fluids, R12, R22, and NH$_3$, with Properties in Accordance with Figure 2

Line no.	Parameter	R12	R22	NH$_3$	Units	Equation no.
1	h_1	181.70	176.34	109.32	(kJ kg^{-1})	
2	h_3	228.62	236.69	338.52	(kJ kg^{-1})	
3	h_4	364.94	413.38	1485.3	(kJ kg^{-1})	
4	h_6	343.48	397.04	1436.6	(kJ kg^{-1})	
5	s_1	0.9308	0.9108	0.6567	(kJ kg^{-1} K$_{-1}$)	
6	s_3	1.0982	1.1254	1.4770	(kJ kg^{-1} K^{-1})	
7	s_4	1.5479	1.7087	5.2602	(kJ kg^{-1} K^{-1})	
8	s_6	1.5699	1.7828	5.8999	(kJ kg^{-1} K^{-1})	
9	h_5	371.71	437.02	1702.8	(kJ kg^{-1})	
10	T_5	39.6	82	109.7	(°C)	
11	h_6-h_3	114.86	160.38	1098.1	(kJ kg^{-1})	
12	h_5-h_6	28.23	39.95	266.2	(kJ kg^{-1})	
13[a]	$\eta(=COP)$	2.85	2.82	2.89	—	3.46
14	h_3-h_1	46.92	60.35	229.20	(kJ kg^{-1})	
15	s_3-s_1	0.1674	0.2146	0.8203	(kJ kg^{-1} K^{-1})	
16	h_5-h_4	6.77	23.54	217.5	(kJ kg^{-1})	
17	s_5-s_4	0.0220	0.0741	0.6397	(kJ kg^{-1} K^{-1})	
18	$w_0(T_1, T_2)$	22.7	31.7	217.0	(kJ kg^{-1} s)	3.19
		(100)	(100)	(100)	(%)	
19[b]	$w_1 = T_2\delta s_1$	5.4	7.1	25.5	(kJ kg^{-1} s)	3.19
	w_1/w_0	(23.8)	(22.4)	(11.7)	(%)	
20[b]	$w_2 = T_2\delta s_2$	0.1	1.0	23.4	(kJ kg^{-1} s)	3.19
	w_2/w_0	(0.4)	(3.1)	(10.8)	(%)	
21	$w_0 + w_1 + w_2$	28.2	39.8	265.9	(kJ kg^{-1} s)	3.19
	$\Sigma w_i/w_0$	(124.2)	(125.5)	(122.5)	(%)	
22[c]	$\eta(=COP)$	2.85	2.82	2.89	—	3.27
23[d]	g_f	80.4	79.6	81.6	(%)	3.22
24[d]	$g(T_1, T_2)$	56.3	55.7	57.2	(%)	3.21
25[d]	$g_{HP}(T_1,T_2)$	63.5	63.0	64.1	(%)	3.30
26[d]	$g(t_1, t_2)$	39.4	39.0	40.0	(%)	3.9
27[d]	$g_{HP}(t_1, t_2)$	46.8	46.5	47.4	(%)	3.31

Note: Input parameters: $T_2 = 30°C$; $T_1 = -20°C$; $\eta(T_1, T_2) = 5.06$; $T_2 - t_2 = 7$ K; $t_1 - T_1 = 7$ K; $g_c = 0.7$.

[a] Enthalpy balance.
[b] Irreversibilities, δs_1 and δs_2.
[c] Entropy balance.
[d] Quality factors for fluids, g_f; for the refrigeration, $g(T_1, T_2)$ and $g(t_1, t_2)$; or for the heat pump process, $g_{HP}(T_1, T_2)$ and $g_{HP}(t_1, t_2)$ and $g_{HP}(t_1, t_2)$.

The quality factor for heat pumping g_{HP} is somewhat larger than that for refrigeration g. For reversible processes, both g and g_{HP} are, by definition, equal to one. For a strongly irreversible process (w finite, $q_1 \to 0$), the factor g approaches zero, whereas g_{HP} stays finite due to electric resistance heating.

3.7.3 Numerical Results
In Tables 3.3, 3.6, and 3.7, results for five typical fluids are shown: H$_2$O and NH$_3$ belong to the group of fluids shown in Figure 3.27; R114 belongs to the group in Figure 3.28.

Rows 11 and 12 in the tables are the enthalpy differences by which the COP (row 13) can be calculated using Equation (3.2). Rows 14 to 17 are enthalpy and entropy differences, from which the entropy production δs_1 and δs_2 are determined (rows 19 and 20). For NH$_3$ in Table 3.5, the additional work due to throttling and desuperheating are both about 11% of the minimum work, w_0. For R22 and R12, the contribution of desuperheating is rather small, if not negligible. Throttling causes about 23% additional work. This is about the same as the sum $w_1 + w_2$ for NH_3. For R114, the term δs_2 is exactly zero.

The total work in row 21 of the tables should be equal to $h_5 - h_6$ in line 12. The small discrepancies are within the accuracy of data charts. Row 22 represents the COP as calculated by an entropy balance. The values agree with those of row 13, as they must.

Row 23 represents the quality factor for the fluids, whereas rows 24 and 25 characterize the quality of the total process, yet without heat transfer. The latter is taken into account in the quality factors of rows 26 and 27. The fluids by themselves reach about 80% of the Carnot efficiency. It is the compressor and the heat transfer which reduce the COP to about 55% and 40%, respectively, of the value for the reversible process. It is interesting to note that the quality factor for water, taken for 50°C temperature lift, has a maximum of 88% as a function of temperature, which is above any value which can be reached with a halogenated refrigerant. The work w_1 is only 5% of w_0 below 100°C and reaches 14% at $T_1 = 250$°C, whereas the work w_2 decreases from 29% to 6% and increases again close to T_c. It should be emphasized that, in Table 3.7, processes with an identical temperature lift of 50°C and not identical pressure ratio or Carnot factor are compared.

In Table 3.8, the work w for NH_3 (see Table 3.5) is divided into different contributions according to Equations 3.19, 3.14, and 3.6. These tables clearly demonstrate how large the contributions of the compression and the heat transfer are compared to those of the fluid. The first numbers in the last rows represent the quality factors g, $g(T_1, T_2)$ and $g(t_1,t_2)$ (see Equations 3.22, 3.19, and 3.11 and column 5 in Table 3.5).

Equations 3.26 and 3.27 have the following advantage compared to the standard Equations 3.6 and 3.7: with the latter equations, the COP is determined by the differences of large numbers which, therefore, must be known with higher precision than required for the COP. With Equations 3.26 and 3.27, the COP is predominantly determined by the Carnot factor, $\eta(T_1,T_2)$, which, for single component fluids, is precisely known once the internal temperatures are chosen. The fluid dependent denominator in Equation 3.27 is a correction of approximately 25% for $T_2 - T_1 = 50$ K (see Tables 3.5 to 3.8), and, therefore, the accuracy requirements are reduced. The absolute value of the heat of evaporation is of minor relevance for the COP. Only ratios of fluid properties, such as $\delta s_1/r$, with the heat of evaporation in the denominator, enter into Equation 3.27. This will become even more evident when analytic equations for the COP are derived. Furthermore, Equation 3.27 shows that the COP is only weakly dependent on the absolute values of the pressures of the individual fluids. These need to be known only with such a precision that the type of compressor can be chosen. The isentropic efficiency for the compressor must be measured separately anyway.

3.7.4 Analytic Equations for the COPs

Analytic equations have the advantage that the dependence and sensitivity on fluid parameters become even more noticeable. Furthermore, fast calculations or estimates are possible. The enthalpy and entropy differences in Equations 3.23 to 3.25 can be calculated using the mean specific heat capacities c and c_p. The following relationships hold:

$$q_1 = r - c\left(T_2 - T_1\right) \tag{3.32}$$

$$T_2\delta s_1 = cT_2\left(\frac{T_2 - T_1}{T_1} - \ln\frac{T_2}{T_1}\right) + v\left(p_2 - p_1\right)\frac{T_2}{T_1}$$

$$\approx \frac{c}{2}\frac{\left(T_2 - T_1\right)^2}{T_1} \tag{3.33}$$

Table 3.6   Parameters for R114 with Properties in Accordance with Figure 3

Line no.	Parameter	R114		Units	Equation no.
a	$T_2$30	70		(°C)	
b	T_1	-20	20	(°C)	
c	$n_{???}(T)$	5.06	5.86	—	
d	g_c	0.70	0.70	—	
e	p_2	2.52	7.43	(bar)	
f	p_1	0.37	1.82	(bar)	
1	h_1	180.59	220.45	(kJ kg^{-1})	
2	h_3	231.01	275.30	(kJ kg^{-1})	
3	h_4	357.51	382.38	(kJ kg^{-1})	
4	h_6	325.18	351.06	(kJ kg^{-1})	
5	s_1	0.9263	1.0719	(kJ kg^{-1} K^{-1})	
6	s_3	1.1072	1.22429	(kJ kg^{-1} K^{-1})	
7	s_4	1.5245	1.5553	(kJ kg^{-1} K^{-1})	
8	s_6	1.4975	1.5175	(kJ kg^{-1} K^{-1})	
9	h_5	349.2	369.41	(kJ kg^{-1})	
10	T_5	30	70	(°C)	
11	h_6-h_3	944.12	75.86	(kJ kg^{-1})	
12	h_5-h_6	24.0	18.35	(kJ kg^{-1})	
13[a]	n (=COP)	2.75	2.89	—	3.46
14	h_3-h_1	50.42	54.75	(kJ kg^{-1})	
15	s_3-s_1	0.1809	0.1710	(kJ kg^{-1} K^{-1})	
16	h_5-h_4	-8.300	-12.97	(kJ kg^{-1})	
17	s_5-s_4	-0.027	-0.038	(kJ kg^{-1} K^{-1})	
18	$w_0(T_1, T_2)$	18.6	12.95	(kJ kg^{-1} s)	3.19
		(100)	(100)	(%)	
19[b]	$w_1 = T_2\delta s_1$	5.45	5.32	(kJ kg^{-1} s)	3.19
	w_1/w_0	(29.3)	(41.1)	(%)	
20[b]	$w_2 = T_2 ds_2$	—	—	(kJ kg^{-1} s)	3.19
	w_2/w_0	(—)	(—)	(%)	
21	$w_0 + w_1 + w_2$	24.05	18.27	(kJ kg^{-1} s)	3.19
	$\Sigma w_i/w_0$	(129.3)	(141.1)	(%)	
22[c]	n (=COP)	2.75	2.89	(%)	3.27
23[d]	g_f	77.3	71.0	(%)	3.22
24[d]	$g(T_1, T_2)$	54.0	49.7	(%)	3.21
25[d]	$g_{HP}(T_1, T_2)$	61.6	57.1	(%)	3.30

[a] Enthalpy balance

[b] Irreversibilities, δs_1 and δs_1

[c] Entropy balance.

[d] Quality factors for the fluid, g_f; for the refrigeration, $d(T_1, T_2)$; and the heat pump process, $g_{HP}(T_1, T_2)$.

v   = specific volume of the liquified working fluid

v_g = specific volume of the gaseous working fluid

In the second term in Equation 3.33, the pressure difference $p_2 - p_1$ has been replaced by $(T_2 - T_1)r/Tv_g$ according to the Clausius-Clapeyron equation. For all practical purposes, this term can be neglected against the first one, except for very small temperature lifts namely for $(T_2 - T_1)/T_2 < 2$ $v/v_g - r/cT$. For typical fluids, the right-hand side of this inequality is about 10^{-2} to 10^{-3}. For such small temperature differences T_2 $\delta s/q_1$ is of the order of 10^{-4} to 10^{-6} and thus negligible anyhow. The entropy production δs_2 can be approximated by:

Table 3.7   Parameters for H₂O

Line no.	Parameter	H₂O					
a	T_1 (°C)	0	50	100	150	200	250
b	T_2 (°C)	50	100	150	200	250	300
c	η_r	5.46	6.47	7.46	8.46	9.46	10.46
d	g_c	0.7	0.7	0.7	0.7	0.7	0.7
e	p_1 (bar)	6×10^{-3}	0.123	1.01	4.76	15.55	39.78
f	p_2 (bar)	123×10^{-3}	1.013	4.76	15.55	39.78	85.92
1	η_1	0.00	209.3	419.1	632.2	852.4	1085.7
2	h_3	209.3	419.1	632.2	852.2	1085.7	1344.9
3	h_4	2592	2676	2746	2793	2801	2749
4	h_6	2501	2592	2676	2746	2793	2801
5	s_1	0.00	0.7038	1.3071	1.8418	2.3308	2.7934
6	s_3	0.7038	1.3071	1.8418	2.3308	2.7934	3.2548
7	s_4	8.0753	7.3547	6.8383	6.4318	6.0721	5.7049
8	s_6	9.1544	8.0753	7.3547	6.8383	6.4318	6.0721
9	h_5	3062	2999	2994	3000	3000	3014
10	T_5	293	263	267	286	317	350
11	$h_6 - h_3$	2291.7	2173	2052.8	1893.6	1707.3	1456
12	$h_5 - h_6$	561	407	318	254	207	213
13[a]	η (=COP)	2.86	3.74	4.52	5.22	5.77	4.78
14	$h_3 - h_1$	209.1	209.8	2213.1	220.2	233.3	259.2
15	$s_3 - s_1$	0.7038	0.6033	0.5347	0.4890	0.4626	0.4614
16	$h_5 - h_4$	470	323	243	207	199	265
17	$s_5 - s_4$	1.0791	0.7206	0.5164	0.4065	0.35997	0.3673
18	w_0	414.0	336.0	275.2	223.8	180.1	139.1
		(100%)	(100%)	(100%)	(100%)	(100%)	(100%)
19[b]	$w_1 = T_2 \delta s_1$	20.1	16.8	14.0	15.1	15.4	20.0
	w_1/w_0	(4.9%)	(5.0%)	(5.1%)	(6.7%)	(8.5%)	(14.4%)
20[b]	$w_2 = T_2 \delta s_2$	121.8	55.0	29.5	15.0	11.0	55.0
	w_2/w_0	(29.2%)	(16.8%)	(10.8%)	(6.7%)	(6.1%)	(39.6%)
21	$w_0 + w_1 + w_2$	555.9	407.8	318.7	253.9	206.5	214.1
	$\Sigma w_i/w_0$	(134.1%)	(121.8%)	(115.9%)	(113.4%)	(114.7%)	(154%)
22[c]	η (=COP)	2.88	3.74	4.50	5.22	5.78	4.76
23[d]	g_f	74.6	82.0	86.3	88.0	87.2	64.9
24[d]	$g (T_1, T_2)$	52.2	57.4	60.4	61.5	61.2	45.4
25[d]	$g_{HP} (T_1, T_2)$	59.6	63.1	65.1	65.6	64.9	50.2

[a] Enthalpy balance.
[b] Irreversibilities, δs_1 and δs_2.
[c] Entropy balance.
[d] Quality factors, g_f, $g (T_1, T_2)$ and $g_{HP} (T_1, T_2)$.

$$T_2 \delta s_2 = c_p T_2 \left(\frac{T_5 - T_2}{T_2} - \ln \frac{T_5}{T_2} \right)$$

$$\approx \frac{c_p}{2} \left(\frac{T_5 - T_2}{T_5} \right)^2 \tag{3.34}$$

The approximations in Equations 3.33 and 3.34 are not valid close to the critical point. We find for the COP:

$$\eta = g_c \eta_r \left\{ 1 + \frac{\dfrac{1}{2\eta_r} \left[\dfrac{cT_1}{r} + \dfrac{c_p T_1}{r} \left(\dfrac{T_5 - T_2}{T_2 - T_1} \right)^2 \dfrac{T_1}{T_5} \right]}{1 - \dfrac{cT_1}{r\eta_r}} \right\}^{-1} \tag{3.35}$$

Table 3.8 Work per Kilogram for a NH$_3$ Compressor System According to Equations 3.19, 3.14 and 3.6, Using the Input Data of Table 3.5

Work s^{-1} =	(Minimum	+Throttle	+Desuperheating	+Compressor	+Condenser	+Evaporator)
Equation 3.19						
w (kW kg^{-1} s)	= 0.7^{-1} [q$_1$/η_r (T)	+T$_2\delta$s$_1$	+T$_2\delta$s$_2$]			
380.0	= 1.42 (217.0	+25.4	+23.4)			
122.5%	= 100%	+11.7%	+10.8%			
100%	= 81.6%	+ 9.6%	+8.8%			
Equation 3.14						
w (kW kg^{-1} s)	= [q$_1$/η_r (T)	+T$_2\delta$s$_1$	+T$_2\delta$s$_2$	+T$_2\delta$s$_3$]		
380.0	= 217.0	+25.5	+23.4	+114.1		
175.1%	= 100%	+11.7%	+10.8%	+52.6%		
100%	= 57.1%	+6.7%	+6.2%	+30.0%		
Equation 3.6						
w (kW kg^{-1} s)	= [q$_1$/η_r (t)	+t$_2\delta$s$_1$	+t$_2\delta$s$_2$	+t$_2\delta$s$_3$	+t$_2\delta$s$_4$	+t$_2\delta$s$_5$]
380.0	= 152.4	+24.9	+22.8	+111.2	+34.6	+34.1
249.3%	= 100%	+16.3%	+14.9%	+73.0%	+22.7%	+22.4%
100%	= 40.0%	+6.5%	+6.0%	+29.3%	+9.1%	+9.0%

The efficiency and quality factors are, in general, determined by three nondimensional fluid parameters (see Tables 3.9 and 3.10):

$$\frac{cT_1}{r}, \frac{c_pT_1}{r}, \frac{T_5-T_2}{T_2-T_1}\left(=\frac{\text{superheating}}{\text{temperature lift}}\right) \tag{3.36}$$

For not too large temperature lifts, $T_2 - T_1$, the superheating is proportional to $T_2 - T_1$:

$$\frac{T_5-T_2}{T_2-T_1} = \left(\frac{r\beta}{c_p}-1\right) \tag{3.37}$$

where β is the thermal expansion coefficient, with $\beta = 1/T$ for ideal gases. Therefore, in this case, the thermal expansion may be used as a third fluid parameter. Further, the temperature dependence of the heat of vaporization

$$\frac{T}{r}\frac{dr}{dT} = -\left(\frac{cT}{r}-\frac{c_pT}{r}\right)-(\beta T-1) \tag{3.38}$$

may be used to characterize the superheating of a fluid, inserting β into Equation 3.37. Table 3.9 gives numerical values for five fluids.

From this table, trends in the physical properties of fluids may be recognized. The larger the values in line 8 of Table 3.9, the larger the entropy production in throttling. The larger the value in line 11, the larger the entropy production from desuperheating. For negative values, this entropy production is zero. The value $r\beta/c_p > 1$ separates superheating fluids of Figure 3.27 from those of Figure 3.28. Table 3.9 may not be used to calculate the superheating of NH$_3$ and especially of H$_2$O quantitatively if $T_2 - T_1 > 50$. For these fluids, the proportionality of Equation 3.37 is not valid for $T_2 - T_1 \approx 50°C$. In the following, we will concentrate on halogenated refrigerants for which, according to Table 1, the contribution of δs$_2$ is zero or can be neglected.

Equations 3.35 and 3.36 have the following interesting property: for small temperature lifts, that is for large values of η_r, the efficiency of all fluids approaches the Carnot efficiency η_r

Table 3.9   Physical Properties and Non-Dimensional Parameters for R114, R12, R22, NH$_3$ and H$_2$O

Line no.		R114		R12	R22	NH$_3$	H$_2$O	
1	T (°C)	50	0	0	0	0	0	125
2	r (kJ kg^{-1})	117.45	138.08	152.54	204.93	1260.7	2501	2182
3	c (kJ kg^{-1} K^{-1})	1.11	1.00	0.93	1.20	4.59	4.20	4.26
4	c_p (kJ kg^{-1} K^{-1})	0.81	0.71	0.63	0.67	2.66	1.90	2.22
5	$\bar{c}_p$ (kJ kg^{-1} K^{-1})	0.85	0.76	0.71	0.81	2.76	1.90	2.12
6	10^{3}β (K^{-1})	4.50	4.07	4.80	5.07	4.83	3.58	2.86
7	(T/r) dr/dT	−1.32	−0.69	−0.88	−1.13	−0.78	−0.26	−0.51
8	cT/r	3.03	1.99	1.75	1.60	1.09	0.46	0.78
9	c_pT/r	2.22	1.42	1.12	0.89	0.63	0.21	0.41
10	βT − 1	0.46	0.13	0.31	0.38	0.32	0.00	0.14
11	(rβ/c_p) − 1	−0.39	−0.21	0.16	0.56	1.09	3.82	1.81
12[a]	(T/r) dr/dT	−1.29	−0.70	−0.85	−1.09	−0.74	−0.25	−0.49

[a] Calculated using Equation 3.37.

Table 3.10   Specific Heat Capacity of the Liquid, c (kJ kg^{-1} K^{-1}), Heat of Evaporation, r (kJ kg^{-1}), and Dimensionless Parameter, cT/r, for R12, R22, and R114

T(°C)	−40	−30	−20	−10	0	10	20	30	40
R12									
c			0.904	0.914	0.926	0.944	0.960	0.982	1.014
r			161.78	157.28	152.54	147.51	142.13	136.32	129.98
cT/r			1.409	1.529	1.658	1.812	1.980	2.184	2.443
R22									
c	1.088	1.126	1.158	1.182	1.204	1.216	1.230	1.242	1.262
r	234.16	227.75	220.73	213.12	204.93	196.16	186.80	176.79	165.97
cT/r	1.0835	1.202	1.328	1.460	1.605	1.756	1.931	2.130	2.382
R114									
c			0.942	0.970	0.996	1.020	1.046	1.066	1.086
r			144.59	141.45	138.08	134.48	130.61	126.50	122.11
cT/r			1.650	1.805	1.971	2.148	2.348	2.555	2.785

T(°C)	50	60	70	80	90	100	110	120	130
R12									
c	1.052	1.102	1.170	1.266	1.416	1.712			
r	122.96	115.07	104.01	91.09	81.34	63.02			
cT/r	2.703	3.191	3.797	4.600	6.243	9.826			
R22									
c	1.300	1.376	1.520	1.824					
r	154.08	140.58	124.49	103.63					
cT/r	2.727	3.261	4.190	6.217					
R114									
c	1.104	1.120	1.138	1.158	1.184	1.220	1.276	1.366	1.544
r	117.45	112.49	107.18	101.47	95.23	88.28	80.28	70.64	58.08
cT/r	3.038	3.317	3.644	4.031	4.516	5.157	6.091	7.603	10.719

multiplied by the compressor efficiency g_c. The quality factor of the fluids g_f approaches 100% (see also Figure 3.30).

For fluids with the properties of Figure 3.28, the contribution δs$_2$ is exactly zero. For R22, the contribution from superheating amounts to only 2.5%; for R12, it amounts to only 0.3% at T$_2$ −T$_1$ = 50$_o$C. Thus, the efficiency for practically all halogenated hydrocarbons is determined by only one parameter, that is, cT$_1$/r, with c = [c(T$_1$ + c(T$_2$)]/2 and r = r(T$_1$).

With $\delta s_2 \approx 0$, Equation 3.35 can be written as:

$$\eta = g_c \eta_r \frac{1 - \dfrac{cT_1}{r}\dfrac{1}{\eta_r}}{1 - \dfrac{1}{2}\dfrac{cT_1}{r}\dfrac{1}{\eta_r}} \tag{3.39}$$

η approaches zero for $cT_1/r = \eta_r$, which is equivalent to $c(T_2 - T_1) = r$. In this case, point 7 (Figures 3.27 and 3.28) coincides with point 6.

For $\eta_r \gg cT_1/2r$, one finds a very simple equation for the COP of a compression refrigerator:

$$\eta \approx g_c \left(\eta_r - \frac{1}{2}\frac{cT_1}{r} \right) \tag{3.40}$$

The Carnot efficiency, η_r, is lowered by a certain value, which is fluid dependent, and then multiplied with the compressor efficiency. In Table 3.10, the values of $c(T)$, $r(T)$ and cT/r are listed for three fluids.

In Figure 3.29, calculated COPs for three fluids are plotted as a function of temperature lift using Equation 3.39 and the values of Table 3.10 for $c(T_1)$, $c(T_2)$, and $r(T_1)$. For R12 and R114, Equation 3.39 yields the COP with the accuracy of existing data charts compared to that calculated by enthalpy balances. For R22, the value of Equation 3.39 is about 3% too large because δs_2 has been neglected. This is not noticeable on the scale of Figure 3.29. Since the temperature dependence of $c(T)$ is rather small, except close to the critical point, practically identical curves are obtained if the value of $c(T_1)$, that is the parameter shown in the third row in Table 6, is used instead of the average specific heat for cT/r in Equation 3.39. Using this value and with $T_2 - T_1 \le 50$, the efficiency is, for most cases overestimated by less than 2%. In Figure 3.30, the quality factor, g_f, for R114 is shown as a function of temperature lift, $T_2 - T_1$. The parameter to the five curves is the evaporator temperature T_1. Starting from 100%, the fluid quality factor, g_f, drops linearly as a function of the temperature lift with a slope $c/2r(T_1)$. At a certain lift, depending on T_1, g_f reaches zero, that is, for $h_3(=h_7) = h_6$.

3.7.5 Conclusions and Comparison with the Exergy Concept

Rather simple equations for the COP of compressor heat pumps and refrigerators have been derived. The fluid properties are characterized by a small set of nondimensional parameters. For new fluids of the halogenated hydrocarbon type, only cT_1/r needs to be determined. The same statement holds for most product fluids of distillation columns, for which vapor recompression is under consideration. Experimentally, the ratio c:r may be determined by a throttling experiment, measuring the ratio of vapor to liquid after throttling. Since c_p and c are roughly proportional to each other, a large value of c:r is also indicative of a nonsuperheating fluid.

Highly fluorinated long-chain halocarbons are considered as potential high temperature working fluids. Due to the large numbers of internal vibration modes of such molecules, the ratio cT:r will be larger. Therefore, it can be expected that such long-chain fluids have reduced efficiency, except if the expansion valve is replaced by an expansion turbine or the liquid is precooled by gaseous fluid.

Exergy (or availability) is the potential work which could be gained if, for example, heat of a certain temperature t was converted into work in a reversible power station working against the lower temperature t_0. Exergy losses are taken as criteria for the performance of a process step. This exergy loss is proportional to the entropy increase δs_i, multiplied by t_0.

The choice of t_0 as ambient temperature is rather ambiguous, and t_0 may be dependent on many circumstances. It changes as a function of parameters which, in general, have no or minor influence on the process performance.

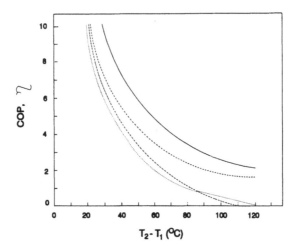

Figure 3.29 Efficiency of three fluids for T = 0°C according to Equation 3.38.
————, η_r (Carnot);   ············, $\eta_c \eta_r$;   –––––––, R12/R22;   ··········· , R114.

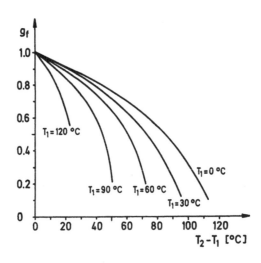

Figure 3.30 Quality factor for R114 as a function of temperature lift $T_2 - T_1$.

It is unsatisfactory that the criteria for performance of a process should depend on an undefined environmental state. The loss of exergy is of relevance for a power station. In this case, this quantity represents per definition the loss of product, which is power output. However, already for a cogeneration plant, the loss in power is not identical with the rate in exergy loss, even if the exergy of the drained heating flow is taken into account. For heat pumps or refrigerators, the merits of the exergy concept are even more unclear. An exergy (availability) balance does not present more information than contained in Equations 3.2 and 3.3, as can be shown as follows: multiplying Equation 3.3 by the parameter t_0 and combining Equations 3.2 and 3.3 yields the exergy balance for the compressor heat pump or refrigerator:

$$w = q_2 \frac{t_2 - t_0}{t_2} + q_1 \frac{t_0 - t_1}{t_1} + t_0 \sum_{i=1}^{n} \delta s_i \qquad (3.41)$$

However, each term on the right-hand side now depends on an unspecified parameter t_0. What can be deduced from the knowledge about the exergy loss $t_0 \Sigma \delta s_i$, that is, the loss in potential work which could be extracted from the condenser and evaporator heat in fictitious reversible power stations, operating against or at temperature level t_0, which can be chosen ambiguously? Of more relevance are the following questions:

1. How much additional work must be supplied to overcome the irreversibilities and to run the process at a given refrigeration or heating power?

2. How much is the product output (that is, the refrigeration duty, q_1, or the heating duty, q_2) at a given power input reduced due to irreversibilities?

The answer to both questions can be found in Equations 3.6, 3.14, 3.19, or 3.28, free of any undefined parameter. The additional work or the reduction in refrigeration power is proportional to entropy production, but the proportionality factors, t_2 or T_2, in general, differ widely from the ambient temperature, t_0. With $T_2 = 150°C$ for a high temperature heat pump, this work, $T_2 \Sigma \delta s_i$, is about 50% higher than the corresponding exergy losses. In "thermoeconomics" the increase of operating costs due to irreversibilities is set proportional to the exergy losses. For heat pumps and refrigerators, this is certainly not correct. The quality or effectiveness of a process can only depend on process properties, that is, on process temperatures and not on environmental temperatures which do not influence the process. For selected plants, such as a thermal power station or an air-liquefaction plant, the ambient happens to be a process temperature, and, therefore, the exergy is useful. For most other cases, such as distillation plants, desalination plants, cogeneration stations, heat pumps, and refrigerators, temperatures other than the ambient are relevant. In Equations 3.9 or 3.21, for the quality factors the entropy increase is multiplied by a temperature factor, $T_1 T_2/(T_2 - T_1)$, which has nothing to do with the ambient temperature. Since for a Second Law analysis by entropy balances no further assumptions are required, these are more basic than a Second Law analysis by exergy balances. Entropy balances can easily be converted into exergy balances, if this should be of advantage. In the past, plant analysis indeed has sometimes been based on the comparison of the entropy production in selected process steps. Yet, as has been shown in this chapter, for heat pumps and refrigerators the individual contributions, δs_i, are not independent from each other. This feature is quite general and can be found, for example, in cogeneration plants, in distillation, or in many other processes in chemical engineering. Therefore, in addition to the absolute values δs_i, the interdependence of the individual contributions must be known before decisions are taken, which may result in costs for additional investment or for changes of the process. The global entropy analysis yields these interdependencies and, in particular, the "currency exchange factors" by which the entropy increases have to be multiplied for conversion to such units, which bridge the connection to economy. Irreversibilities can then be measured, either in prices of the work or of the heat needed to run the process, or in prices of the product delivered by the process.

3.7.6 Appendix

The following relationships hold for entropy and enthalpy balances along the path 1–3–4–5–6–1 in Figures 3.27 and 3.28:

$$\frac{r_1}{T_1} - \frac{r_2}{T_2} = s_3 - s_1 + s_5 - s_4 \tag{3.42}$$

$$r_1 - r_2 = h_3 - h_1 + h_5 - h_4 - \left(h_5 - h_6\right) \tag{3.43}$$

$$r_1 = q_1 + h_3 - h_1 \tag{3.44}$$

Eliminating r_1 and r_2 from Equations 3.42, 3.43, and 3.44 and inserting $h_5 - h_6$ into the following equations:

$$w = \frac{1}{g_c}\left(h_5 - h_6\right) \qquad (3.45)$$

$$\eta = g_c \frac{h_6 - h_3}{h_5 - h_6} \qquad (3.46)$$

yields Equations 3.26 and 3.27. By the entropy balance Equation 3.42, the Second Law of Thermodynamics is added onto the enthalpy balance, and, thus, the fluid independent Carnot factor can be extracted. The more detailed derivation given in the above text has the advantage of greater clarity. The individual terms are easier to interpret, and more complicated modifications of the Rankine process, such as fluid injection or liquid expansion, can be handled, knowing the contribution of each entropy source separately.

REFERENCES

1. Ahlby, L., Hodgett, D. L., Berntsson, T. (1991): Optimization study of the compression/absorption cycle. In: *International Journal of Refrigeration,* Vol. 14, No. 1, P. 16-23.

2. Alefeld, G. (1983): Design optimization for multistage compression, expansion and absorption devices. In: Proceedings of the 16th Int. Congress of Refrigeration, Paris, paper no. A3-157.

3. Altenkirch, E. und Tenckhoff, B., (1911b): Absorptionskältemaschine zur kontinuierlichen Erzeugung von Kälte und Wärme oder auch von Arbeit. Deutsches Reichspatent Nr. 278076.

4. Altenkirch, E. (1951): Der Einfluß endlicher Temperaturdifferenzen auf die Betriebskosten von Kompressionskälteanlagen mit und ohne Lösungskreislauf. In: Kältetechnik, Bd. 3, Nr. 8, S. 201, Nr. 9, S. 229, Nr. 10, S. 255.

5. Bercescu, V. e.a. (1983): Aspects du fonctionnement d'une installation experimentale de pompe de chaleur avec compression mecanique et circulation additionelle de la solution. Preprints to the 16th International Congress of Refrigeration, Paris, Com. E2, paper no. 585, p. 173.

6. Bergmann, G., Hivessy, G. (1987): Hybrid heat pump for hot water supply. In: Proceedings of the 17th Int. Congress of Refrigeration, Wien, vol. E, p. 250.

7. Bergmann, G., Hivessy, G. (1988a): The Multi-Stage Hybrid Heat Pump. In: Proceedings of the 2nd Int. Workshop on Research Activities on Advanced Heat Pumps, Graz.

8. Bergmann, G., Hivessy, G. (1988b): Development of hybrid heat pumps. In: Proceedings of the 2nd Int. Workshop on Research Activities on Advanced Heat Pumps, Graz.

9. Bergmann, G., Hivessy, G. (1989): Hybrid heat pump pilot plant. In: Proceedings of the IEA Heat Pump Center Workshop "High Temperature Heat Pumps", November 1989, Hannover. Report No. HPC-WR-5, IEA Heat Pump Center, Karlsruhe.

10. Bergmann, G. (1990): Efficiency of wet compression in hybrid heat pumps. In: Proceedings of the 3rd Int. Workshop on Research Activities on Advanced Heat pumps, Graz.

11. Bergmann, G., Hivessy, G. (1990): Experimental hybrid heat pump of 1000 kW heating capacity. In: Proceedings of the 4th Int. Conference on Application and Efficiency of Heat Pump Systems in Environmentally Sensitive Times, Oct. 1-3, Munich, Germany. Sti, Oxford.

12. Brunnaben, K. (1964): Die Kombination von Kompressions — und Absorptions-Kälteanlage zur Wirtschaftlichen Ausnutzung von hochgespanntem Dampf. In: Kältetechnick, Jg. 16, no. 7, S. 211.

13. Chiriac, F. (1987): The use of absorption heat pumps as heat recovery plants. In: Proceedings of the 17th Int. Congress of Refrigeration, Wien, vol. B., p. 1153.

14. Costa, A. (1991): Betrieb einer Kompressions-Absorptions- Wärmepumpe. Diplomarbeit am Institut für Technische Physik E 19, Physik-Department der Technischen Universität München.

15. Dietrich, E., Le Goff, P. (1986): Un nouveau type de pompe á chaleur: la PAC á compression-absorption a fonctionnement alterné. In: Revue Physique Appliquée, vol. 21, p. 45-52.

16. Friedel, W. e.a. (1986): Heat pumps for heat recovery from paper dryers, producing process steam from the dryer exhaust air. Final report No. BF-R-65.044-1, Commission of the European Communities, Contract No. EE-B 1-152-D (B).

17. George, J. M., Marx, W., Srinivasa Murthy, S. (1990): A comparative thermodynamic study of R22-DMETEG and R22-DMF compression-absorption heat pumps. In: Heat Recovery Systems & CHP, vol. 10, no. 1, p 31-36.

18. Hämmer, G. (1990): Eine Kompressions-Absorptionskältemaschine mit hohem Temperaturhub. Diplomarbeit am Institut für Technische Physik E 19, Physik-Department der Technischen Universität München.

19. Hodgett, D. L., Friedel, W. (1982): Heat pumps for heat recovery from paper dryers, producing process steam from the dryer exhaust air. In: New Ways to Save Energy. Report of the CEC, EUR 8077N, p. 165.

20. Hodgett, D. L., Ahlby, L. (1987): The effect of the properties of the refrigerant and solvent on the compression/absorption cycle. In: Proceedings of the 17th Int. Congress of Refrigeration, Wien, vol. B, p.1018.

21. Hodgett, D. L., Ahlby, L. (1988): Compression-absorption heat pumps. In: Absorption heat pumps. Proceedings of a workshop held in London, 12.–14. April 12–14. Report No. EUR 11888EN, Commission of the European Communities, p. 204-215.

22. Jeday, M.R., LeGoff, P. (1987): La pompe á chaleur á compression-absorption. Influence des différences de prix de l'électricité nuit/jour sur le coût d'exploitation. In: Revue Physique Appliquée, vol. 22, p. 445-456.

23. Kawada, A., Otake, M., Toyofuku, M., Ota, H. (1988): Absorption heat transformer with TFE/E181 pair. Advanced Absorption Workshop, October. (SS-ACHP)

24. Lourdudoss, S., Stymne, H. (1985): About the possibility of using mineral oil-fluorocarbon refrigerant working pair in a compressor heat pump with solution cycle. In: Proceedings of the Int. Workshop on Heat Transformation and Storage, Ispra, Italy. Report No. S.A./I.04.D2.85.35., Commission of the European Community, Joint Research Center, Ispra Establishment.

25. Malewski, W. F. (1988a): Integrated absorption and compression heat pump cycle using mixed working fluid ammonia and water. Proceedings of the Inst. of Refrigeration 1987-88, 4-1. London 1988.

26. Malewski, W. F. (1988b): Integrated absorption and compression heat pump cycle using mixed working fluid ammonia and water. In: Proceedings of the 2nd International Workshop on Research Activities on Advanced Heat Pumps, Graz.

27. Morawetz, E. (1986): Sorptions-Kompressionsvärmepumpar och Värmetransformatiorer. Statens rad för byggnadsforskning, Stockholm, R38: 1986 (In schwedischer Sprache).

28. Morawetz, E. (1989): Sorption-compression heat pumps. In: *International Journal of Energy Research*, vol. 13, p. 83.

29. Mucic, V. (1984): A new method for COP-increasing of two-media resortion compression heat pumps with solution circuit. In: VDI-Verlag, Düsseldorf.

30. Mucic, V. (1988): Resorption Compression Heat Pump with Solution Circuit for Steam Generation Using Waste Heat of Industry as Heat Source. Paper presented at the Absorption Experts Meeting 1988, Dallas.

31. Mucic, V. (1989): Resorption compression heat pump with solution circuit for steam generation using waste heat of industry as Heat Source. In: Proceedings of the IEA Heat Pump Center, vol. 7, no. 1, p. 14-17

32. Mucic, V. (1990): Compression-absorption heat pumps In: Heat Pumps. Solving Energy and Environmental Challenges. *Proceedings of the 3rd International Energy Agency Heat Pump Conference*, Tokyo, Japan, March 12-15. 1990, Pergamon Press, Oxford, p. 391-400.

33. Najork, H. (1991): Resorption cycles. In: Refrigeration Cycles with Emphasis on Environmental Aspects and Energetic Efficiency. Report of the working group of the International Institute of Refrigeration (IIR), presented at the 1991 IIR Congress in Montreal. Ch. 4.2

34. Niebergall, W. (1955): Thermische zusammenschaltung von Kompressions-und Absorptions-Kältemaschine. In: Allgemeine Wärmetecvhnik, Jg. 6, Nr. 8/9, S. 161-169; Jg. 7, Nr. 1,S. 1-9.

35. Novotny, S. (1979): Possibilities of improving the thermodynamic working conditions of heat pumps for heat recovery. In: *International Journal of Refrigeration*, vol.2, p. 171.

36. Osenbrück, A. (1985): Verfahren zur Kälteerzeugung bei Absorptionsmaschinen. Deutsches Reichspatent DRP 84084.

37. Ostermayer, S. (1989): Aufbau und Simulation einer Absorptions-Kompressions-Kältemaschine. Diplomarbeit am Institut für Technische Physik E 19, Physik-Department der Technischen Universität München.

38. Otake, M., Kawada, A., Toyofuku, M., Ota H. (1990): Heat Storage and Heat Pump System using TFE/E181. Paper presented at the IEA Annex XVI workshop, March 1990.

39. Pourreza-Djourshari, S., Radermacher, R. (1986): Calculation of the performance of vapour compression heat pumps with solution circuits using the mixture R22-DEGDME. In: *International Journal of Refrigeration*, vol. 9, p. 245.

40. Pritchard, C., Low, R. (1988): A self-regulating heat pump for systems with variable power input. In: Proceedings of the 2nd Int. Workshop on Research Activities on Advanced Heat Pumps, Graz.

41. Radermacher, R. (1987): Vapour compression heat pump cycle with desorber/absorber heat exchange. In: Proceedings of the 17th Int. Congress of Refrigeration, Wien, vol. B, p. 1061.

42. Radermacher, R. (1988): Manipulation of vapor pressure curves by thermodynamic cycles. *Transactions of the ASME, Journal of Engineering for Gas Turbines and Power*, vol. 110, p. 647.

43. Radermacher, R., Howe, L. (1988): Internal Combustion Engine Driven Heat Pump Cycles. Paper presented at the Absorption Experts Meeting, Dallas.

44. Radermacher, R., Herold, K. E., Howe, L. A. (1988): Combined vapor compression/absorption cycles. In: Absorption Heat Pumps. Proceedings of a workshop held in London, April, 12–14. Report No. EUR 11888EN, Commission of the European Communities, p. 225.234.

45. Radermacher, R., Zheng, J., Herold, K. E. (1988): Vapor compression heat pump with two stage solution circuit: proof-of-concept unit. In: Absorption Heat Pumps. Proceedings of a workshop held in London, April, 12–14. Report No. EUR 11888EN, Commission of the European Communities, p. 204-224.

46. Rane, M.V., Radermacher, R., Herold, K. E. (1989): Experimental Investigation of a Single Stage Vapor Compression Heat Pump with Solution Circuit. Paper presented at the Winter Annual Meeting of the ASME, San Francisco.

47. Riffat, S. B. (1988): A thermochemical/compressor heat pump using 'economy 7' tariff electricity. In: *International Journal of Refrigeration*, vol. 12, p. 521-526.

48. Scheuermann, B., Mucic, V. (1984): 2-Stoff-Kompressions-Wärmepumpe mit Lösungskreislauf. Versuchsanlage Mannheim-Waldhof. BMFT Forschungsbericht BMFT-FB-T 84-197.

49. Schwarzhuber, J. (1989): Aufbau und Betrieb einer Kompressions-Absorptions-Kältemaschine. Diplomarbeit, am Institu für Technische Physik E 19, Physik-Department der Technischen Universitä München.

50. Sellerio, U. (1957): Machines Frigorifiques á Absorption-Compression. Bulletin de l'Institut international du Froid, Annexe. Bd. 2, S. 131. Zitiert nach Niebergall (1981), S. 162.

51. Spindler, U. (1990): Betrieb einer Kompressionskältemaschine mit integriertem Absorptionskreis. Diplomarbeit am Institut für Technische Physik E 19, Physik-Department der Technischen Universität München.

52. Stokar, M. R., von Neuforn (1986): Kompressionswärmepumpe mit Lösungskreislauf. Dissertation Nr. 8101, Eidgenössische Technische Hochschule Zürich.

53. Stokar, M., Trepp, C. (1987): Compression heat pump with solution circuit. Par 1: design and experimental results. In: *International Journal of Refrigeration*, vol. 10, p. 87.

54. Sunye, R., Prevost, M., Bugarel, R. (1988): High temperature sorption cycle for heat pumping: Compressor aided heat transformer. In: Absorption Heat Pumps. Proceedings of a workshop held in London, April, 12–14. Report No. EUR 11888EN, Commission of the European Communities, p. 197-203.

55. Voigt, H. (1980): Wärmepumpe und Verfahren zu ihrem Betrieb. Patentschrift DE 3044580 C2.

56. Voigt, H. (1982): Kätemaschinenprozesse für Speicherbetrieb. In: KK - Die Kälte und Klimatechnik, Jg. 35, Nr. 8, S. 312-318.

57. Voigt, H. (1985): Heat Pumping and Transforming Processes with Intrinsic Storage. In: Energy Conversion Management, vol. 25, no. 3, p. 381-386.

58. Ziegler, F. (1988): Advanced Compression Absorption Cycles. Paper presented at the Absorption Experts Meeting, Dallas.

59. Ziegler, F., Alefeld, G. (1989): Compression absorption Heat Pumps for High Temperature Applications, In: Proceedings of the IEA Heat Pump Center Workshop "High Temperature Heat Pumps", November 1989, Hannover. Report No. HPC-WR-5, IEA Heat Pump Center, Karlsruhe.

60. Ziegler, F. (1991) Kompressions/Absorptions Wärmepumper, Dissertation, Techinsche Universitat Munchen.

61. (For Section 3.7) Alefeld, G. (1987a): Efficiency of compressor heat pumps and refrigerators derived from the second law of thermodynamics. In: *International Journal of Refrigeration,* vol. 10, no. 11, p. 331.

62. (For Section 3.7) Alefeld, G. (1989): Second law analysis for an absorption chiller. In: *Newsletter of the IEA Heat Pump Center,* vol. 7, No.2 (June), pp. 54-57.

CHAPTER FOUR

Absorption Heat Pumps (Refrigerators) and Heat Transformers

This chapter discusses single-stage absorption heat pumps and heat transformers. State-of-the-art concepts are explained, some implementations are introduced, and examples for the evaluation of absorption systems are given.

4.1 PRINCIPLE OF OPERATION

The essential difference between absorption- and vapor-compression heat pumps is that the absorption system converts heat of a given temperature to heat of another temperature without any intermediate use of work. The effect of an absorption heat pump can be explained using Figure 4.1. Heat is transferred reversibly from T_2 to T_1, i.e., by releasing its content of availability, and the waste heat, Q_1', is rejected at T_1. The availability is then used within the cycle to pump heat Q_1'' from T_0 to T_1. For the vapor-compression heat pump, the output and input of availability in the form of work are conducted separately in the power plant and at the consumer's location, respectively. In the following, it will be shown how an absorption heat pump avoids the detour of employing mechanical work.

4.2 THE ABSORPTION HEAT PUMP, COMBINATION OF HEAT ENGINE AND HEAT PUMP

Figure 4.2 shows that a vapor-compression heat pump working between T_0 and T_1 can be powered by a work-producing cycle operating between T_2 and T_1. Both cycles may employ different working fluids. The requirement that the generated work is completely consumed by the heat pump results in a relationship governing the two circulating working fluid streams. If both cycles use the same working fluid and if they contain solution circuits, and if, further, the low and high pressures in both cycles are the same, then a compressor and a turbine can be eliminated and the two pairs of exchange units can be connected directly. The result is the machine shown in Figure 4.3, representing an absorption heat pump. Without incorporating the detour of generating work, the cycle in Figure 4.3 can be explained using the concept derived in Chapter 3.6. Within the first pair of exchange units, A'B', the availability of heat is converted into the availability of a high pressure gas. This gas passes through a second set of exchange units in which the availability of the high pressure gas is converted back into availability of heat. The temperature levels at which heat is supplied or rejected in both pairs of exchange units differ when the concentrations of the absorbent fluid are different. The temperature levels are shifted by applying the physical phenomenon of "boiling point elevation" or "vapor pressure reduction". The exchange units A and B, Figure 4.3, may contain pure working fluid only. In this case, the solution circuit between A and B degenerates to one line containing either a throttle or a pump, depending on the direction of the circulation.

If the overall direction of flow of the working fluid is counterclockwise, then Figure 4.3 represents an absorption heat pump or refrigerator. By supplying heat at the highest temperature level (driving heat), refrigeration is provided at the lowest temperature level, and heat is rejected at the two intermediate temperature levels. If the working fluid circulates clockwise, then Figure 4.3 represents a heat transformer. Heat is supplied at the two intermediate

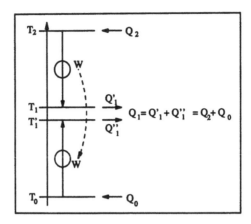

Figure 4.1 The flow of heat and work in an absorption heat pump.

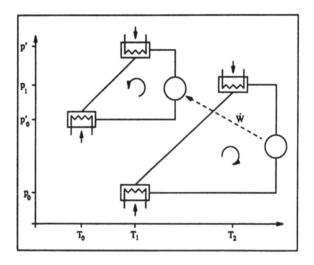

Figure 4.2 Vapor-compression heat pump powered by a power generation cycle.

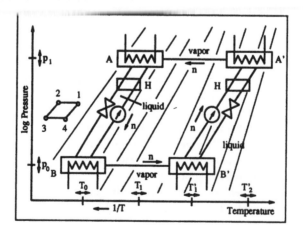

Figure 4.3 Schematic diagram of an absorption heat pump cycle in $\ln(P),(-1)/T$.

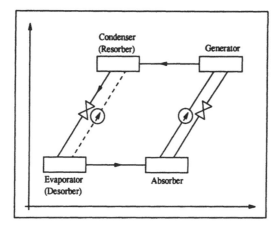

Figure 4.4 Absorption heat pump cycle. The dashed line with the pump can be omitted.

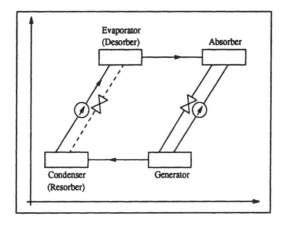

Figure 4.5 Absorption heat transformer cycle. The dashed line with the expansion valve can be omitted.

temperature levels, and heat is rejected at both the lowest and highest temperature levels. Figures 4.4 and 4.5 state the common notation for the components. If the pair of exchange units operating at the lower temperature level contains only pure working fluid, the pipe with the solution pump (Figure 4.4) or the expansion device (Figure 4.5) can be omitted. The notations in parenthesis are valid for those units which have two solution circuits. Historically they are called "resorbers" and "desorbers", although the processes are identical to those in absorbers and generators.

4.3 DESCRIPTION OF THE COMPLETE ABSORPTION CYCLE

The cycle employed in an absorption heat pump includes the processes that are shown in detail in Figure 4.6. The absorbent is component A and the refrigerant is component B of a mixture AB:

1   Out of the solution AB vapor B is evaporated by means of the driving heat Q_2 (generator),

1–2 Vapor B is transferred at constant pressure p_1 to the condenser,

2   Vapor B condenses (condenser) at T_1, releasing heat Q_1'',

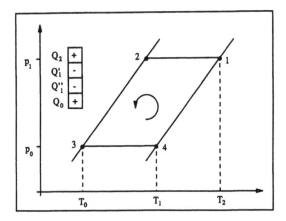

Figure 4.6 Processes in an absorption heat pump.

2–3   Pressure is reduced by expansion to P_0, and the liquid B is cooled to T_0,

3   Liquid B evaporates, absorbing the heat Q_0 (evaporator),

3–4   Vapor B is transferred at constant pressure P_0 to the absorber,

4   Vapor B is absorbed by A, forming a solution AB rejecting the heat Q_1 at T_1 (absorber),

4–1   The pressure of the solution AB is increased from P_0 to P_1, and AB is transferred from the absorber to the generator. Simultaneously, fresh absorbent is transferred from the generator to the absorber.

For processes 1 and 3, heat is supplied to the cycle while it is rejected by processes 2 and 4. Processes 2 (condensation) and 3 (evaporation) are the same as those in a vapor-compression heat pump.

Figure 4.7 is a three-dimensional plot of the individual process steps occurring in an absorption heat pump. For the sake of simplicity, the change of concentration is assumed to be so small that the strong and weak absorbents are represented by one vapor pressure curve. In this chapter, the following terminology will be used: Solutions which are rich in working fluid are called weak absorbent; solutions which are poor in working fluid are called strong absorbent. These definitions coincide with the common use of strong and weak salt solutions (e.g., lithiumbromide water), but are in contrast to the common terminology for ammonia water.

Both Rankine Sorption Cycles are combined so that the process steps in the expansion turbine (4–6′) of the high temperature cycle and those in the compressor (6′–4) of the low temperature cycle cancel each other. Both Rankine Sorption Cycles must operate between the same pressure levels P_2 and P_1. The low temperature cycle may operate without an absorbent.

In Figure 4.7 and in its projections, Figure 4.8, the following processes can be recognized:

1–2   Increase in pressure for the weak absorbent,

2–3   Pressurization and heating of the solution,

3–4   Desorption,

4–4′   Desuperheating at constant pressure,

4′–3′   Condensation,

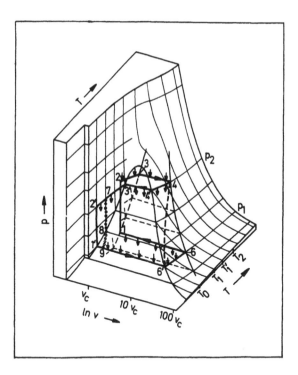

Figure 4.7 The absorption heat pump (or absorption heat transformer) as a combination of two Rankine Sorption Cycles.

3′–7 Cooling of the working fluid,

7–9 Isenthalpic expansion,

9–6′ Evaporation,

6′–6 Superheating at constant pressure,

6–1 Absorption,

4–6 Cooling and depressurization of the strong absorbent.

The diagram showing the components in Figure 4.8h can be understood as a combination of the diagrams of Figures 3.2h and 3.3h, Chapter Three, by eliminating the turbine and compressor, respectively.

The T,s-diagram, Figure 4.8d, exhibits projections of two surfaces, one for the pure fluid and one for the solution. The cycle operates clockwise (generation of work) on the first. It operates counterclockwise (input of work) on the second.

The technical realization of the single-stage absorption heat pump results in two circulation loops, one for the working fluid and one for the absorbent, as shown in Figure 4.9. The two loops are clockwise and counterclockwise and share a common line, the line containing the weak absorbent. In general terms, the reactions occurring within an absorption heat pump can be written as follows:

$$A\,B \text{ (solid, liquid)} + Q_2 \rightarrow A \text{ (solid, liquid)} + B \text{ (vapor)}$$

$$B \text{ (vapor)} \rightarrow B \text{ (liquid)} + Q_1' \text{ (pressure } P_1)$$

$$B \text{ (liquid)} + Q_0 \rightarrow B \text{ (vapor) (pressure } P_0)$$

$$B \text{ (vapor)} + A \text{ (solid, liquid)} \rightarrow AB \text{ (solid, liquid)} + Q_1'' \tag{4.1}$$

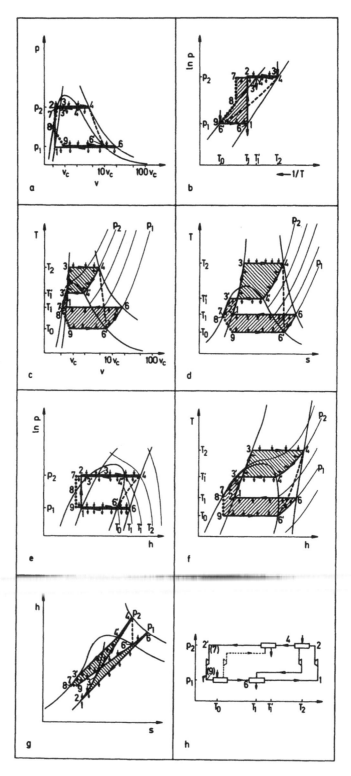

Figure 4.8 Projections of a work-generating Rankine Sorption Cycle combined with a work-consuming Rankine Cycle (absorption heat pump). When the directions of all arrows are inverted, both cycles exchange their tasks with regard to production or consumption of work (heat transformer).

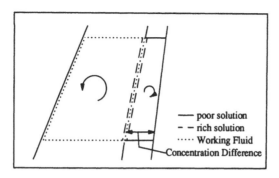

Figure 4.9 Schematic diagram of fluid circulation in an absorption heat pump.

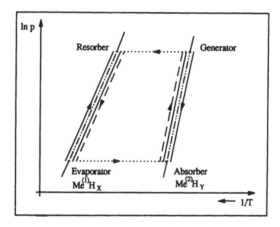

Figure 4.10 Absorption heat pump with hydrogen as refrigerant.

Symbol A on the right-hand side may represent a pure substance or a compound or a solution still containing the component B. Examples for these reactions are shown in Figures 3.10 and 3.11, Chapter Three.

The vapor pressure curves for H_2O/NH_3 and $H_2O/LiBr-H_2O$ are shown in Figures 2.7 and 2.15, Chapter Two. Since crystallization of the salt would disturb the operation of a continuously working unit, the solubility limit of the salt is of special interest (see Figure 2.15). Whereas the pressures in system NH_3/H_2O above 30°C evaporator temperature are always higher than 1 bar, the evaporator pressures are in the vacuum-range bar for single-stage LiBr/H_2O systems. On the other hand, system NH_3/H_2O produces pressures above 20 bar for heat rejection at temperatures above 50°C in the condenser. A further disadvantage is the toxicity of NH_3. System $LiBr/CH_3OH$ has pressure levels which are higher by a factor of three compared with $LiBr/H_2O$. However, the toxicity and poor thermal stability of CH_3OH above 120°C in the presence of salt hamper its use.

Pumping of solid adsorbents is not a simple matter. Therefore, the reactions in Figures 3.10 and 3.11 using solid adsorbents require a different technology. This is discussed in more detail in Chapter Eleven. Here, only the solid-vapor systems are briefly considered as examples for resorbers. An absorption cycle using hydrogen as a working fluid can be built without liquifying H_2. Two metal hydrogen alloys with different vapor pressure lines are used as absorber and resorber, respectively. The liquid phase of H_2 is, therefore, replaced by the solid system Me^1H_x (Figure 4.10) which has, at any temperature, a higher vapor pressure than does system $Me^{(2)}H_y$. A resorption unit employing liquid adsorbents shows three circulation loops (Figure 4.11) for heat pump operation.

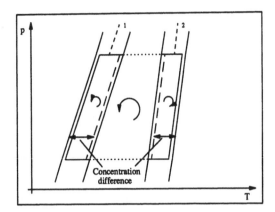

Figure 4.11 Resorption cycle with three circulation loops.

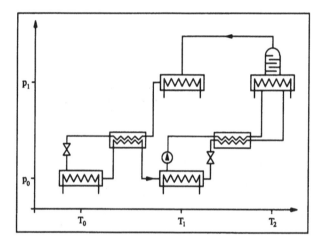

Figure 4.12 Detailed schematic diagram of an absorption heat pump.

Figure 4.12 shows a schematic representation of an absorption heat pump in more detail. In addition, two internal heat exchangers are shown. One is used for heat exchange within the solution circuit. This heat exchanger is a requirement for achieving a good cycle efficiency. It may be designed as a three-path heat exchanger, including in the third path the gas line between generator and condenser. The second heat exchanger is used to precool the liquid phase of the working fluid before entering the expansion valve and to preheat the vapor of the working fluid leaving the evaporator.

Figure 4.13 shows a schematic diagram of a $LiBr/H_2O$ water chiller. Because of the low vapor pressure of water, gas lines must be avoided. The pairs of exchange units, generator and condenser and evaporator and absorber, are combined into two shells. There are also designs which incorporate all four exchange units into one hermetically sealed shell, Figure 4.14.

4.4 COEFFICIENTS OF PERFORMANCE

The coefficients of performance (heat ratios) can be calculated on different levels of sophistication, depending on the required precision and availability of fluid data. In this chapter, first the theoretical heat ratios which can be achieved with reversibly operating machines are discussed. A more precise method using specific heat values and latent heat only is described in Chapter 8.4.1. If complete enthalpy-concentration charts are available, the heat ratios can

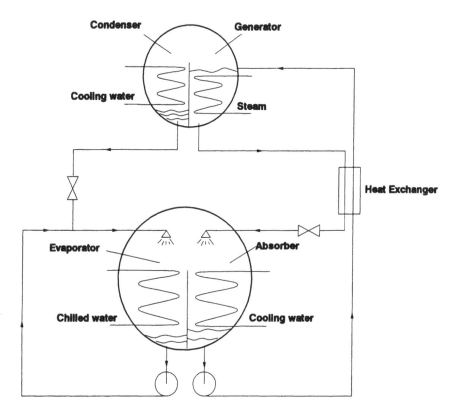

Figure 4.13 Process scheme of a single-stage LiBr-water chiller with two shells.

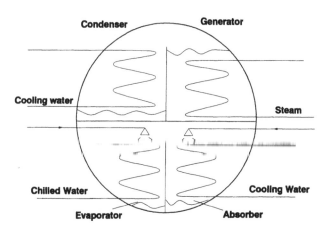

Figure 4.14 Process scheme of a single-stage LiBr-water chiller in one shell.

be calculated by applying enthalpy balances for individual components of the cycle. This application of this method is demonstrated in Example 4.1. In Chapter 4.8, a method for evaluating and analyzing cycles using, in addition to enthalpy balances, the Second Law (entropy balances) explicitly is presented.

For the various operating modes of an absorption cycle, the following COPs can be derived for reversible machines, as shown in Chapter One:

(a) Absorption refrigerator with a cooling capacity of Q_0 and a generator heat input of Q_2:

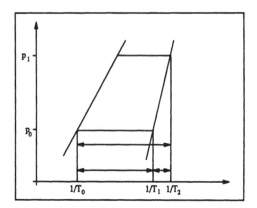

Figure 4.15 Heat ratios expressed as ratios of distances on an inverse temperature scale.

$$\xi_1 = \frac{Q_0}{Q_2} = \frac{\dfrac{1}{T_1} - \dfrac{1}{T_2}}{\dfrac{1}{T_0} - \dfrac{1}{T_1}} = \frac{\dfrac{T_2 - T_1}{T_2}}{\dfrac{T_1 - T_0}{T_0}}$$ (4.2)

(b) Absorption heat pump with a heating temperature level T_1, (heat is pumped from T_0 to T_1):

$$\xi_2 = \frac{Q_1}{Q_2} = \frac{\dfrac{1}{T_0} - \dfrac{1}{T_2}}{\dfrac{1}{T_0} - \dfrac{1}{T_1}} = \frac{\dfrac{T_2 - T_0}{T_2}}{\dfrac{T_1 - T_0}{T_1}}$$ (4.3)

(c) Absorption heat pump with heating temperature level T_2, (heat is pumped from T_1 to T_2) (heat transformer):

$$\xi_3 = \frac{1}{\xi_2} = \frac{Q_2}{Q_1} = \frac{\dfrac{1}{T_0} - \dfrac{1}{T_1}}{\dfrac{1}{T_0} - \dfrac{1}{T_2}} = \frac{\dfrac{T_1 - T_0}{T_1}}{\dfrac{T_2 - T_0}{T_2}}$$ (4.4)

These equations are valid independent of a particular technical implementation of the heat pump. As the equations demonstrate, the efficiencies can be obtained as the ratio of differences in $1/T$ taken from the vapor pressure curves in Figure 4.15. This figure represents the cycle of a single-stage absorption heat pump according to the schematic diagram in Figure 4.3. Therefore, it can be expected that the efficiencies ξ_1, ξ_2, and ξ_3 can be expressed by the thermodynamic properties of the working fluid pair. Clapeyron's equation for a pure component is given as:

$$\frac{dp}{dT} = \frac{s_2 - s_1}{v_2 - v_1} = \frac{r}{T(v_2 - v_1)}$$ (4.5)

with $r = T(s_2 - S_1)$ representing the latent heat.

Far from the critical region ($v_1 \ll v_2$) and by assuming the validity of the ideal gas law $Pv_2 = RT$, Clapeyron's equation can be simplified to:

$$-R\frac{d\left(\ln(p)\right)}{d\left(\dfrac{1}{T}\right)} = r \tag{4.6}$$

With these assumptions, the slope of the vapor pressure curve in a $\ln(p)$-$1/T$ diagram is proportional to the heat of evaporation.

An analogous equation can be derived for the vapor pressure over a solution:

$$-R\frac{d\left(\ln(p)\right)}{d\left(\dfrac{1}{T}\right)} = r + 1 \tag{4.7}$$

The change in the heat of evaporation corresponds to the heat of mixing, l, of the condensed vapor in the solution. The following relationship between the various amounts of heat transferred to and from the exchange units of a single-stage absorption unit can be derived:

$Q_0 = Q_1' = r$

$Q_2 = Q_1'' = r + 1$

$Q_1 = Q_1' + Q_1'' = 2r + 1$

With these relationships, the following theoretical COP's or heat ratios are obtained:

$$\xi_1 = \frac{Q_0}{Q_2} = \frac{r}{r+1}$$

$$\xi_2 = \frac{Q_1}{Q_2} = \frac{2r+1}{r+1} \tag{4.8}$$

$$\xi_3 = \frac{Q_2}{Q_1} = \frac{r+1}{2r+1}$$

Considering the finite specific heat of the working fluid and of the absorbent, corrections have to be applied, which will be discussed in Chapter 8.5.

Three cases can be distinguished with regard to the COP when the heat of mixing, l, is taken into account. They are represented in Figures 4.16 and 4.17. The efficiencies of a single-stage reversible absorption cycle are limited in certain ways, as shown in Figure 4.17. An absorption heat pump with a working fluid mixture showing a positive heat of mixing can never have a COP higher than 2. This statement is surprising, remembering that an absorption heat pump can be seen as composed of a heat engine and a heat pump, as shown in Figure 4.2. Yet the inherent coupling of the two cycles, as displayed in Figure 4.3, requires that the same working fluid flow rates circulate through both devices, at identical pressure levels. These requirements couple temperature levels. Choosing T_1 and T_0, T_2 is fixed and can only be shifted slightly by selecting a working pair with a different l.

The discussion of the COP as a function of the heat of solution may be helpful, but the thermodynamic reasons are better understood with the following consideration: In Figure 4.18(a), it is assumed that the heat of mixing is positive and that the temperature T_1 is kept constant. The desorber temperature has to be lower compared to the case of nonexisting heat of mixing. The availability of the heat Q_2 with respect to T_1 is also lower. Consequently, more heat Q_2 is needed to pump a given amount of heat from T_0 to T_1. The efficiencies ξ_1 and ξ_2

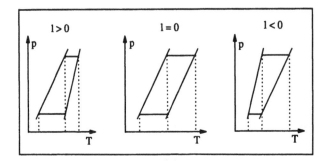

Figure 4.16 Shift of relative location of temperature levels as a function of the heat of mixing.

	Refrigeration $T_0 \rightarrow T_1$	Heat Pump $T_0 \rightarrow T_1$	Heat Transformer $T_1 \rightarrow T_2$
$1 > 0$	$0 < \xi_1 < 1$	$1 < \xi_2 < 2$	$1/2 < \xi_3 < 1$
$1 = 0$	$\xi_1 = 1$	$\xi_2 = 2$	$\xi_3 = 1/2$
$1 < 0$	$1 < \xi_1 < \infty$	$2 < \xi_2 < \infty$	$0 < \xi_3 < 1/2$

Figure 4.17 Ranges of COP in dependence of the sign of the heat of mixing.

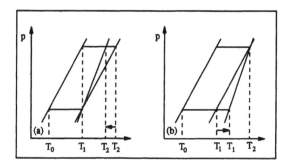

Figure 4.18 Shift of temperature levels in dependence of the heat of solution.

are reduced. However, when temperature T_2 is fixed and still a positive heat of mixing is assumed, the absorber temperature shifts to higher values, Figure 4.18(b). In this case, the increase of the temperature level T_1 at which heat is rejected is accompanied by the reduction of the efficiency. Any change of parameter l, the heat of mixing, changes the availability of the transferred heat and, therefore, the amount of heat required.

Example 4.1: Absorption Heat Pump Evaluation

The following absorption heat pump is to be evaluated. The working fluids are water, used as the refrigerant, and an aqueous solution of lithium bromide, used as the absorbent. The evaporator temperature is 5°C. The lowest temperature in the absorber and the condenser temperature is 40°C. It is assumed that the liquid at the absorber and desorber outlet is saturated. The concentration of strong absorbent is 63%.

(a) What is the solution flow rate through the pump based on 1 kg of refrigerant circulating through the evaporator?

(b) What are the amounts of heat exchanged in the evaporator, condenser, absorber, desorber, and solution heat exchanger based on 1 kg of refrigerant vapor?

(c) What is the COP of the system?

Solution: Figure 4.19 shows a process scheme defining the state points. State point 1 represents the liquid water in the evaporator; its temperature is 5°C. The saturation pressure, according to Figure 2.15 (Chapter Two), is 0.87 kPa and the liquid enthalpy is 21 kJ/kg. The vapor in equilibrium with the liquid water, point 1′, has an enthalpy of 2510 kJ/kg according to Figure 3.4. The condenser pressure is 7.4 kPa, and the liquid enthalpy is 168 kJ/kg, point 2. The concentration of the solution leaving the absorber is 0.578, point 3. This value is determined from Figure 2.15 at the intersection of the isobar of 0.87 kPa with the isotherm of 40°C. The enthalpy value, 106 kJ/kg is found in Figure 2.1 (Chapter Two) at the intersection of the isostere of 0.578 with the isotherm of 40°C. A schematic diagram is shown in Figure 4.20. The desorber outlet, point 5, has a temperature of 95°C. This can be found in Figure 2.15 at the intersection of the condenser pressure with the isostere of 0.63. The enthalpy, 232 kJ/kg, is found in Figure 4.20 at the intersection of the isostere of 0.63 with the isotherm of 86°C. Point 4 is the beginning of the desorption process. It is located in Figure 2.15 at the intersection of the high pressure isobar with the isostere of 0.578. The temperature is 80°C. The enthalpy, 207 kJ/kg, is found as indicated in Figure 4.20 at the intersection of the 80°C isotherm with the isostere of 0.578. Point 6 is the beginning of the absorption process. Its temperature, 51°C, is again determined in Figure 2.15, where the low pressure isobar intersects with the isostere of 0.63. The enthalpy, 154 kJ/kg, is also found, as indicated in Figure 4.20. The vapor leaving the desorber, point 4′, is assumed to have a temperature of 80°C. Its enthalpy for the pressure of 7.4 kPa is found to be 2650 kJ/kg in Figure 3.4, Chapter Three, the T,s diagram for water. State point 3 is located in the superheated vapor region at the intersection of the isobar of 7.4 kPa (0.074 bar which leaves the saturated vapor line at 40°C) with the isotherm of 80°C. The values are summarized in Table 4.1.

(a) The calculations are performed for 1 kg of vapor flowing through the evaporator. The concentration of the vapor leaving the desorber, x_v, is assumed to be 0.0. The vapor phase contains no salt. The flow rate of the solution through the solution pump amounts to:

$$f = (x_v - x_5)/(x_3 - x_5)$$

This results again from the application of the mass balance for the salt in the desorber, Figure 3.23, Chapter Three.

For f, the following value is found:

$$f = (0.0 - 0.63)/(0.578 - 0.63) = 12.1$$

(b) Next, the solution heat exchanger is evaluated. For this purpose, the enthalpies of all the streams entering and leaving have to be determined. The enthalpy of the liquid leaving the pump is $h_{3'} = h_3 + w_p/f$. The pump work amounts to:

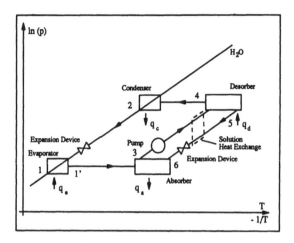

Figure 4.19 Schematic diagram of an absorption heat pump.

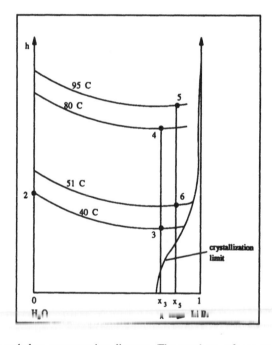

Figure 4.20 Schematic enthalpy-concentration diagram. The numbers refer to state points in Figure 4.19.

Table 4.1 Thermodynamic Properties at the State Points

	T(C)	P(kPa)	h(kJ/kg)	x
1	5	0.87	21	0
1'	5	0.87	2510	0
2	40	7.4	168	0
3	40	0.87	106	0.578
4	81	7.4	189	0.578
4'	81	7.4	2650	0
4''	69	7.4	172	0.578
5	92	7.4	226	0.63
6	51	0.87	154	0.63

$$w_p = f(P_2 - P_1)v_s$$

$$w_p = 12.1(7.4 \text{ kPa} - 0.87 \text{ kPa})\, 1/1200 \text{ m}^3\text{kg} = 0.07 \text{ kJ/kg}$$

where f is the flow rate through the pump, $P_2 - P_1$ is the pressure difference the pump has to overcome, and v_s is the specific volume of the fluid being pumped. Here, an approximate density is assumed. Since the pump work is very small compared to any other amount of energy exchanged in the system, the assumption is acceptable. Actually, the pump work is negligible. Therefore, it is assumed that the enthalpy h_3 does not change. The temperature change due to the pump is also negligible. The fluid entering the desorber can be preheated up to 80°C, point 4. The heat capacity of the stream from 3 to 4, q_p amounts to:

$$q_p = f(h_4 - h_3) = 12.1(189 \text{ kJ/kg} - 106 \text{ kJ/kg}) = 1004 \text{ kJ/kg}$$

The heat capacity of the stream from 5 to 6, q_r amounts to:

$$q_r = (f-1)(h_5 - h_6) = 11.1(226 \text{ kJ/kg} - 154 \text{ kJ/kg}) = 799 \text{ kJ/kg}$$

The amount of heat exchanged is 799 kJ/kg. Thus, the stream from 5 to 6 is cooled to T_6, while the stream from 3 to 4 is not heated to T_4. The enthalpy for the fluid leaving the solution heat exchanger is now calculated based on the requirement that q_p has to be equal to q_r.

$$h_{4''} = h_3 + q_r/f = 106 \text{ kJ/kg} + 799 \text{ kJ/kg}/12.1 = 172 \text{ kJ/kg}$$

The temperature at this point can be read from Figure 2.1, Chapter Two, to 73°C.

Now all enthalpy values are known, and the remaining heats can be calculated. For the evaporator heat q_e, the following value is found:

$$q_e = (h_{1'} - h_2) = (2510 \text{ kJ/kg} - 168 \text{ kJ/kg}) = 2342 \text{ kJ/kg}$$

for the condenser heat q_c:

$$q_c = (h_{4'} - h_2) = (2650 \text{ kJ/kg} - 168 \text{ kJ/kg}) = 2482 \text{ kJ/kg}$$

for the absorber heat q_a (the equation is derived in Example 3.4, Chapter Three):

$$q_a = h_{1'} - h_6 + f(h_6 - h_3)$$

$$q_a = 2510 \text{ kJ/kg} - 154 \text{ kJ/kg} + 12.1(154 \text{ kJ/kg} - 106 \text{ kJ/kg}) = 2937 \text{ kJ/kg}$$

and the desorber heat q_d (Example 3.4):

$$q_d = h_{4'} - h_5 + f(h_5 - h_{4''})$$

$$q_d = 2650 \text{ kJ/kg} - 226 \text{ kJ/kg} + 12.1(226 \text{ kJ/kg} - 172 \text{ kJ/kg}) = 3077 \text{ kJ/kg}$$

Checking the energy balance yields:

$$q_d + q_e - q_c - q_a = 3077 \text{ kJ/kg} + 2342 \text{ kJ/kg} - 2482 \text{ kJ/kg} - 2937/\text{kg} = 0 \text{ kJ/kg}$$

The energy balance is fulfilled.

(c) The COP of the system is

$$\text{COP} = q_e/q_d = 0.76$$

Thus, for each unit of heat supplied to the desorber, the absorption chiller produces 0.76 units of cooling capacity.

Remark: In this evaluation, enthalpy values from different thermodynamic charts are used. As mentioned in Example 3.4, care has to be taken that the reference point for all enthalpies is the same. This is the case with charts and tables used in this example.

4.5 MEANS FOR INCREASING THE COP

Internal Heat Exchangers
The solution heat exchanger shown in Figure 4.12 is, in principle, not necessary for the actual operation of an absorption heat pump, but it is very important for improving efficiencies. Usually there exists a mismatch in the heat capacity of the two absorbent streams in the solution heat exchanger for the following reasons:

A. The mass flow rates are different in the two paths of the heat exchanger (f = weak absorbent flow rate per kilogram of working fluid, Figure 4.21).

B. The specific heat capacity may be a function of composition.

Therefore, in the case of an absorption heat pump, a temperature difference occurs at the high temperature end of the heat exchanger, as shown in Figure 4.21. The weak absorbent enters the desorber without being preheated to the maximum desirable temperature. In the case of a heat transformer, the temperature difference is found at the cold end of the heat exchanger. In this case, the weak absorbent carries valuable high temperature heat from T_2 to T_1, Figure 4.22.

In order to further increase the COP, there are the following additional opportunities:

Three-Pass Heat Exchanger
The vapor leaving the generator is also cooled in the solution heat exchanger. This is now a three-pass heat exchanger. In a heat transformer, the vapor entering the absorber is preheated.

Use of Heat of Rectification
Especially for the working pair NH_3/H_2O, water vapor contained in the gas leaving the generator has to be kept at a minimum. Otherwise, the water eventually accumulates in the evaporator and is unable to contribute to the cooling capacity because of its very low vapor pressure. The water must be transferred from the evaporator back to the absorber as a liquid. In this water, a large amount of ammonia is dissolved, which is lost for the cooling capacity. A reduction of the water content is achieved by rectification of the vapor leaving the generator. The heat generated by condensing the water vapor out of the ammonia/water vapor mixture can be used, for example, by further preheating the weak absorbent. The solution is passed first

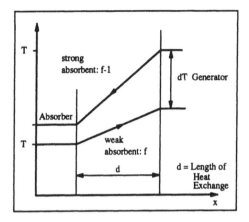

Figure 4.21 Heat exchange between weak and strong absorbent (absorption heat pump).

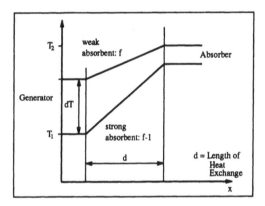

Figure 4.22 Heat exchange between weak and strong absorbent (absorption heat transformer).

through the rectifier to achieve rectification most effectively and then passed through the solution heat exchanger (Figure 4.23).

Example 4.2: Absorption Heat Pump Evaluation with Rectification

An absorption heat pump is to be evaluated. The effect of rectification is included. The working fluid pair is ammonia, used as the refrigerant, and a water/ammonia solution, used as the absorbent. The evaporator temperature is 5°C, the lowest temperature in the absorber and the condenser temperature are 40°C, and the concentration of strong absorbent (solution low in ammonia concentration) is 0.36. Assume that the liquid leaving the absorber and desorber is saturated. The effectiveness of the solution heat exchanger is assumed to be 1.0. The vapor leaving the desorber is rectified to a concentration of 0.995.

(a) What are the pressures, temperatures, and enthalpies of all relevant state points, including the vapor concentration?

(b) What are the amounts of heat exchanged in the evaporator, condenser, absorber, generator, rectifier, and solution heat exchanger per 1 kg of circulating refrigerant vapor?

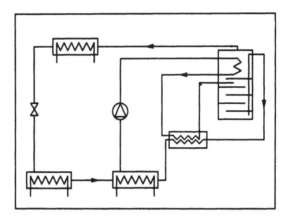

Figure 4.23 Use of the heat of rectification.

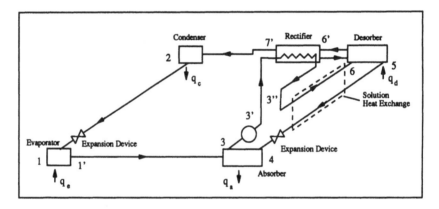

Figure 4.24 Schematic diagram of an absorption heat pump with rectification.

(c) What is the COP of the system?

Solution: (a) The state points are indicated on the process scheme of an ammonia/
water-absorption heat pump in Figure 4.24 and also on Figure 4.25, a
schematic enthalpy-concentration diagram. All property data are sum-
marised in Table 4.2. Point 1' is the ammonia vapor leaving the evaporator. Its pressure is 470 kPa. The pressure is read off the h,x diagram
where the isotherm of 5°C of the liquid region intersects the isostere of
x = 0.995. It is assumed here that the rectifier produces ammonia vapor
of this concentration. For all practical purposes, this is pure ammonia
vapor. However, in Figure 4.25 the vapor concentration is shown as
significantly less than 0.995 to demonstrate the evaluation procedure in
more detail. The enthalpy of the vapor phase is found at the intersection
of the isobar of 470 kPa of the vapor phase with the isostere of x = 0.995.
Since the isobars for high ammonia concentrations are very steep, the
enthalpy is best found by following the auxiliary isobar of 470 kPa. The
value for the vapor enthalpy at point 1' is 1311 kJ/kg. The enthalpy of the
saturated liquid is 22 kJ/kg, point 1. Point 2 is the liquid leaving the
condenser. The pressure, 1520 kPa, is found at the intersection of the
isotherm of 40°C with the isostere of x = 0.995. The enthalpy of the
liquid is 194 kJ/kg. The saturated liquid at the absorber outlet is point 3.

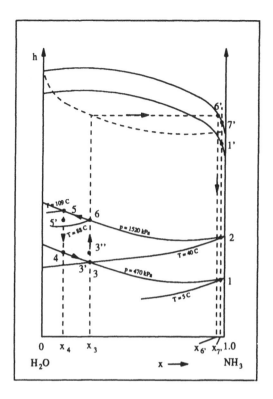

Figure 4.25 Schematic enthalpy-concentration diagram. The numbers of the state points refer to Figure 4.24.

Table 4.2 Thermodynamic Properties at the State Points

	T(C)	P(kPa)	x	h(kJ/kg)
1	5	470	0.995	22
1′	5	470	0.995	1311
2	40	1520	0.995	194
3	40	470	0.46	-63
3″	47	1520	0.46	-34
4	58	470	0.46	31
4′	60	1520	0.36	40
5	109	1520	0.36	285
6	88	1520	0.46	173
6′	88	1520	0.970	1455
7′	40	1520	0.995	1340

It is located at the intersection of the low pressure isobar of 470 kPa and the isotherm of 40°C. The concentration is 0.46, and the enthalpy is –63 kJ/kg. Point 4 is the beginning of the absorption process if the process begins with saturated liquid. It is located at the intersection of the low pressure isobar with the isostere at x = 0.36. The enthalpy is 31 kJ/kg, and the temperature is 58°C. Point 5 represents the end point of the desorption process. It is located at the intersection of the high pressure isobar and the isostere of 0.36. The enthalpy is 285 kJ/kg, and the temperature 109°C. Point 6 is the beginning of the desorption process and is located at the intersection of the high pressure isobar and the isostere of 0.46. The temperature is 88°C, and the enthalpy is 173 kJ/kg. It is assumed that the desorber is constructed in a way that allows the vapor to leave in equilibrium with the solution at the beginning of the desorption process. Following the procedure described in Chapter Two, the vapor concentration is found to be 0.97, and its enthalpy is 1455 kJ/kg, point 6'.

The following assumptions are made about the rectifier: The vapor leaving the rectifier has a concentration of 0.995. It is essentially pure ammonia, point 7'. Its enthalpy is 1340 kJ/kg, found at the intersection of the high pressure isobar with the isostere of x = 0.995. If a better resolution of the graph or computer routines for ammonia/water properties were available, the enthalpy value could be determined more accurately. For the sake of this example, the approach taken here is deemed sufficiently accurate. The solution returning from the rectifier is assumed to be saturated and has a concentration and enthalpy equal to the one at point 6. (The assumptions regarding the rectifier are simplified and may lead to rectifier heats that are smaller than the actual value; a detailed description can be found in the literature.) With this result, all relevant properties of the state points are found.

(b) The evaluation of the cycle follows the same general pattern used in earlier examples, with one exception: the use of the rectifier has to be included. First, the pump work w_p is determined. The flow rate f through the pump is

$$f = (x_7 - x_5)/(x_6 - x_5)$$

$$f = (0.995 - 0.36)/(0.46 - 0.36) = 6.4$$

The pump work amounts to:

$$w_p = f(P_2 - P_6)v = 6.4\,(1520\text{ kPa} - 470\text{ kPa})/800\text{ kgm}^{-3} = 8.4\text{ kJ/kg}$$

With this, the enthalpy $h_{3'}$ of the fluid leaving the pump equals:

$$h_{3'} = h_3 + w_p/f = (-63\text{ kJ/kg}) + 8.4\text{ kJ/kg}/6.4 = -62\text{ kJ/kg}$$

This value is only slightly different from the h_3.

Next, the heat of rectification must be evaluated. The amount of heat rejected by the rectifier is calculated first. The equation for the rectifier heat below is based on an energy and mass balance for the rectifier. The amount of condensate returning to the desorber as reflux v is:

$$v = (x_7 - x_{6'})/(x_{6'} - x_6)$$

$$v = (0.995 - 0.970)/(0.970 - 0.460) = 0.049$$

$$q_r = (h_{6'} - h_7) + v(h_{6'} - h_6)$$

$$q_r = (1455\text{ kJ/kg} - 1340\text{ kJ/kg}) + 0.049\,(1455\text{ kJ/kg} - 173\text{ kJ/kg}) = 178\text{ kJ/kg}$$

The heat of rectification is added to the solution leaving the pump, and its enthalpy increases to $h_{3''} = h_{3'} + q_r/f = -34$ kJ/kg. The temperature of the solution is increased to 47°C, as seen in the h,x diagram in Figure 4.25. Now the solution heat exchanger can be evaluated.

$$q_{rs} = f(h_6 - h_{3^*}) = 6.4\,(173\text{ kJ/kg} - (-34\text{ kJ/kg})) = 1325\text{ kJ/kg}$$

$$q_{ls} = (f - 1)(h_5 - h_4) = 5.4\,(285\text{ kJ/kg} - 31\text{ kJ/kg}) = 1372\text{ kJ/kg}$$

q_{rs} is the smaller amount of heat compared to q_{ls} and the enthalpy of point 4 has to be recalculated.

$$h_{4'} = h_5 - q_{rs}/(f - 1) = 285\text{ kJ/kg} - (1325\text{ kJ/kg})/5.4 = 40\text{ kJ/kg}$$

Now all relevant amounts of heat can be determined. From energy and mass balances for each of the heat exchangers follows:

$$q_e = h_{1'} - h_2 = 1311\text{ kJ/kg} - 194\text{ kJ/kg} = 1117\text{ kJ/kg}$$

$$q_c = h_7 - h_2 = 1340\text{ kJ/kg} - 194\text{ kJ/kg} = 1146\text{ kJ/kg}$$

$$q_a = h_{1'} - h_{4'} + f(h_{4'} - h_3)$$

$$q_a = 1311\text{ kJ/kg} - 40\text{ kJ/kg} + 6.4(40\text{ kJ/kg} - (-63\text{ kJ/kg})) = 1930\text{ kJ/kg}$$

$$q_d = h_{6'} - h_5 + f(h_5 - h_6) + q_r$$

$$q_d = 1340\text{ kJ/kg} - 285\text{ kJ/kg} + 6.4(285\text{ kJ/kg} - 173\text{ kJ/kg}) + 178\text{ kJ/kg} = 1950\text{ kJ/kg}$$

Checking the energy balance yields:

$$q_d + q_e + w_p - q_c - q_a = (1950 + 1117 + 8 - 1146 - 1930)\text{ kJ/kg} = -1\text{ kJ/kg}$$

This result is not exactly equal to zero because of round-off errors. A deviation of less than 1% of the cooling capacity is acceptable.

The COP amounts to:

$$COP = q_e/q_d = 0.57$$

Here, the pump work which also represents an energy input to the heat pump is neglected. Therefore, the complication of determining how to account for the different amounts of availability associated with the desorber heat and the work input to the pump is avoided.

Solution Recirculation

In Figure 4.26, another means for increasing the coefficient of performance is shown. The solution flow is changed such that the solution is preheated or precooled in a heat exchanger inside the absorber or the generator. This method has been termed solution recirculation. The hot, strong absorbent leaving the generator is cooled by rejecting part of its heat to the desorbing mixture in the generator. The cold, weak absorbent leaving the absorber is preheated in the absorber before it enters the solution heat exchanger. For further discussion, it is assumed that the solution heat exchanger is infinitely large. Otherwise, the statements have to be modified slightly.

There are some differences in the effect of the solution recirculation, depending on the application to an absorption heat pump or heat transformer. However, first both systems are considered together (Figures 4.26 and 4.27). If no solution recirculation is used, the strong absorbent will be cooled within the solution heat exchanger to the equilibrium temperature of the weak absorbent leaving the absorber. Thus, subcooled strong absorbent enters the absorber. This lowers the average absorption temperature. When solution recirculation is used in the absorber, then the strong absorbent enters the absorber at its own equilibrium temperature, and the absorption heat is transferred to the sink at the highest possible temperature.

A similar statement holds for the generator. When solution recirculation is employed, the weak solution leaves its high temperature heat content in the generator for further desorption. The use of solution recirculation reduces for the absorption heat pump the temperature difference between the generator and absorber, and for the absorption heat transformer the temperature difference between the generator and absorber is increased. In compression/absorption systems, the temperature lift is reduced, improving the COP.

It has been noted that there exists a mismatch in the heat capacities of the strong and weak absorbents within the solution heat exchanger. In the case of an absorption heat pump without solution recirculation, the weak absorbent usually enters the generator subcooled. The degree of subcooling is determined by the temperature difference between generator and absorber and the difference in the concentration between the weak and strong absorbents. When solution recirculation is employed, the weak absorbent is preheated by the absorber, reducing its degree of subcooling at the generator inlet. Thus, the heat requirement of the generator is reduced, improving the efficiency of the heat pump. This effect does not exist for the heat transformer.

Desorber-Absorber Heat Exchange

To obtain absorber-desorber heat exchange, the difference in concentration of the solution streams entering and leaving the desorber and absorber is chosen to be so large that the highest temperature of the absorber is higher than the lowest temperature of the generator. An internal heat exchange becomes possible. Heat is transferred from the warm end of the absorber to the cold end of the desorber, Figure 4.28. In this case, the solution heat exchanger is not required. When solution recirculation is employed, both solution streams circulate at matching temperatures between the exchange units, as illustrated in Figure 4.29.

The cycle shown in Figure 4.29 has the essential characteristic of a two-stage system which is the heat exchange between a high temperature absorber and a low temperature generator. This is explained in detail in Chapter 7. Other fluid connections with more than one pump between absorber and generator are also shown in Chapter 7.9.

Example 4.3: Absorption Heat Pump with Desorber-Absorber Heat Exchange

An absorption heat pump with desorber-absorber heat exchange and with ammonia and water as the working fluid pair is evaluated. The evaporator temperature is 5°C, the lowest temperature in the absorber and the condenser temperature are 40°C, and the concentration of strong absorbent is 0.10. Saturated fluid is leaving the desorber and absorber. The effectiveness of the desorber-absorber heat exchange is assumed to be 1.0.

(a) What are the pressures, temperatures, and enthalpies of all relevant state points, including the vapor concentration?

(b) What are the amounts of heat exchanged in the evaporator, condenser, absorber, generator, rectifier, and solution heat exchanger per 1 kg of circulating refrigerant vapor?

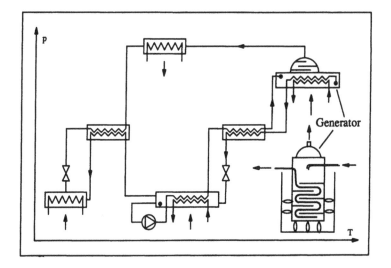

Figure 4.26 Solution recirculation (absorption heat pump).

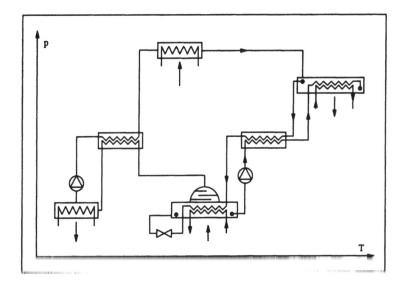

Figure 4.27 Solution recirculation (absorption heat transformer).

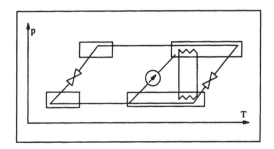

Figure 4.28 Schematic diagram of an absorption heat pump with desorber-absorber heat exchange.

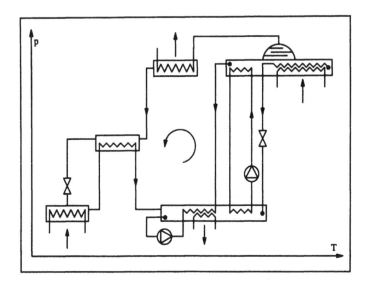

Figure 4.29 Absorption heat pump cycle with desorber-absorber heat exchange and solution recirculation.

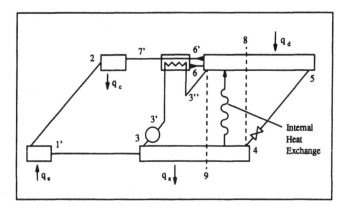

Figure 4.30 Schematic diagram of an absorption heat pump with desorber-absorber heat exchange for Example 4.3.

(c) What is the COP of the system?

Solution: (a) The state points are indicated on the process scheme in Figure 4.30 and also on Figure 4.31, a schematic enthalpy-concentration diagram. All property data are summarized in Table 4.3. Point 1′ is the ammonia vapor leaving the evaporator. Its pressure is 470 kPa. The pressure is read off the h,x diagram where the isotherm of 5°C of the liquid region intersects the isostere for $x = 0.995$. It is assumed here that the rectifier produces ammonia vapor of this concentration. For all practical purposes, this is pure ammonia vapor. The enthalpy of the vapor phase is found at the intersection of the isobar of 470 kPa of the vapor phase with the isostere of $x = 0.995$. Since the isobars for high ammonia concentrations are very steep, the enthalpy is best found by following the auxiliary isobar of 470 kPa. The value for the vapor enthalpy at point 1′ is 1311 kJ/kg, and the enthalpy of the saturated liquid is 22 kJ/kg, point 1. Point 2 is the liquid leaving the condenser. The pressure is found at the intersection of the

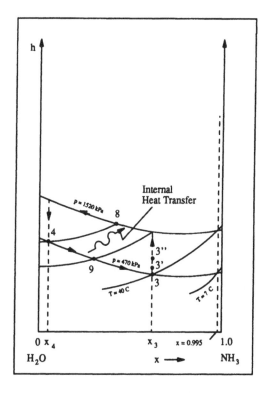

Figure 4.31 Schematic enthalpy-concentration diagram for Example 4.3. The numbers refer to the state points of Figure 4.30.

Table 4.3 Thermodynamic Properties at the State Points

	T(C)	P(kPa)	x	h(kJ/kg)
1	5	470	0.995	22
1'	5	470	0.995	1311
2	40	1520	0.995	194
3	40	470	0.46	-63
3"	47	1520	0.46	9
4	108	470	0.10	386
5	163	1520	0.10	645
6	88	1520	0.46	173
6'	88	1520	0.970	1455
7'	40	1520	0.995	1340
9	88	470	0.20	242
9'	88	470	0.87	1630

isotherm of 40°C with the isostere of x = 0.995, 1520 kPa. The enthalpy of the liquid is 194 kJ/kg. The saturated liquid at the absorber outlet is point 3, located at the intersection of the low pressure isobar of 470 kPa and the isotherm of 40°C. The concentration of the liquid is 0.46, and the enthalpy is –63 kJ/kg. At point 4, the absorption process begins if saturated liquid were used. This point is located at the intersection of the low pressure isobar with the isostere at x = 0.10. The enthalpy is 386 kJ/kg, and the temperature is 108°C. Point 5 represents the end point of the desorption process and is located at the intersection of the high pressure isobar and the isostere of 0.10. The enthalpy is 645 kJ/kg, and the temperature is 165°C. Point 6 is the beginning of the desorption process and is located at the intersection of the high pressure isobar and the isostere of 0.46. The temperature is 88°C, and the enthalpy is 163 kJ/kg. It is assumed that the desorber is constructed in such a way that the leaving vapor is in equilibrium with the solution at the start of the desorption process. Following the procedure described in Chapter Two, the vapor concentration is found to be 0.97, and its enthalpy is 1455 kJ/kg, point 6′.

The vapor leaving the rectifier at point 7′ is assumed to have a concentration of 0.995, which is considered to be pure ammonia. Its enthalpy is 1340 kJ/kg, found at the intersection of the high pressure isobar with the isostere for x = 0.995. The solution returning from the rectifier is assumed to be saturated and to have a concentration to equal the concentration found at point 6. This simplification may lead to an underestimated heat of rectification, but is accurate as long as the difference in concentration between the desorber vapor and rectified vapor is small. The concentrations and enthalpies at points g and 8 have to be calculated as shown in (b) below. In Figure 4.31, the isotherms through points g and 8 are indicated. They bracket the range in which desorber-absorber heat exchange can occur.

(b) The evaluation of the cycle follows the same general pattern used in earlier examples. In addition to the rectifier heat, the desorber-absorber heat exchange has to be taken into account. First, the pump work w_p is determined. The flow rate f through the pump is

$$f = \left(x_\tau - _5\right)/\left(x_6 - x_5\right)$$

$$f = (0.995 - 0.10)/(0.46 - 0.10) = 2.5$$

The pump work is

$$w_p = f\left(P_2 - P_6\right)v = 2.5 \left(1520 \text{ kPa} - 470 \text{ kPa}\right)/800 \text{ kg/m}^{-3} = 3.3 \text{ kJ/kg}$$

With this, the enthalpy of the fluid leaving the pump, $h_{3'}$, equals:

$$h_{3'} = h_3 + w_p/f = (-63 \text{ kJ/kg}) + 3.3 \text{ kJ/kg}/2.5 = -62 \text{ kJ/kg}$$

This value is only slightly different from the h_3.

Next, the heat of rectification has to be evaluated. The amount of heat rejected by the rectifier is calculated according to Example 4.2.

$$v = \left(x_7 - x_{6'}\right)/\left(x_{6'} - x_6\right)$$

$$v = (0.995 - 0.970)/(0.970 - 0.460) = 0.049$$

$$q_r = \left(h_{6'} - h_7\right) + v\left(h_{6'} - h_6\right)$$

$$q_r = (1455 \text{ kJ/kg} - 1340 \text{ kJ/kg}) + 0.049 \, (1455 \text{ kJ/kg} - 173 \text{ kJ/kg}) = 178 \text{ kJ/kg}$$

The heat of rectification is added to the solution leaving the pump, and its enthalpy increases to $h_{3''} = h_{3'} + q_r/f = 9$ kJ/kg. The temperature of the solution increases to 47°C, as can be seen in the h,x diagram in Figure 4.31. The next step is to evaluate the desorber-absorber heat exchange.

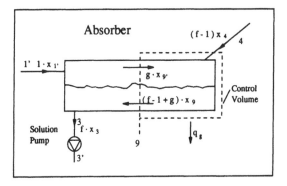

Figure 4.32 Schematic diagram of the absorber.

First, the amount of heat q_g which the absorber supplies to the desorber must be determined. This is the amount of heat that is released at a temperature higher than the lowest temperature of the desorber, 88°C, as seen at point 9 in Figures 4.30 and 4.31. The concentration of the liquid phase at point 9 is 0.20, and its enthalpy is 242 kJ/kg. The vapor at point 9′ is in equilibrium with the liquid at point 9; point 9′ is assumed to have the same temperature. Its concentration is 0.87, and its enthalpy is 1630 kJ/kg. The amount of vapor that is absorbed between points 4 and 9 is designated as g. Figure 4.32 shows a schematic diagram of the absorber. The mass balance yields for a control volume around the high temperature end of the absorber, according to Figure 4.32:

$$(f-1)x_4 + gx_{9'} = (f-1+g)x_9$$

The first term represents the amount of ammonia entering with the liquid stream at point 4 of the control volume. The second term represents the amount of ammonia entering with the vapor stream. The third term represents the amount of ammonia leaving the control volume with the liquid phase. It is assumed that the concentration and enthalpy and, therefore, the temperature of the vapor entering the control volume are in equilibrium with the liquid phase leaving the high temperature section of the absorber.

For g, the following equation is obtained:

$$g = (f-1)(x_9 - x_4)/(x_{9'} - x_9)$$

The validity can be made plausible by checking the limits for g = 1 and g = 0. The numerical value of g is

$$g = 1.5\,(0.20 - 0.10)/(0.87 - 0.20) = 0.224$$

The energy balance yields:

$$q_g = gh_{9'} - h_4 + h_9(1-g) + f(h_4 - h_9)$$

$$q_g = 0.224\,1630 \text{ kJ/kg} - 645 \text{ kJ/kg} + 242 \text{ kJ/kg}\,0.776 + 2.5\,(645 \text{ kJ/kg} - 242 \text{ kJ/kg})$$

$$q_g = 915 \text{ kJ/kg}$$

In the above equation, h_5 was used instead of h_4 because the liquid is not cooled before leaving the desorber. Point 4 denotes the start of the absorption process if the liquid has been cooled to the saturation temperature of 108°C at 470 kPa. This is not the case in this example. However, solution recirculation could be employed, as shown earlier.

The amount of heat released from the absorber to the out side q_{ao} is the total absorber heat minus q_g:

$$q_{ao} = h_{1'} - h_5 + f(h_5 - h_3) - q_g$$

$$q_{ao} = 1311 \text{ kJ/kg} - 645 \text{ kJ/kg} + 2.5 \left(645 \text{ kJ/kg} + -63 \text{ kJ/kg}\right) - 915 \text{ kJ/kg}$$

$$q_{ao} = 1521 \text{ kJ/kg}$$

The net desorber heat amounts to:

$$q_{do} = h_7 - h_5 + f(h_5 - h_{3''}) + q_r - q_g$$

$$q_{do} = 1340 \text{ kJ/kg} - 645 \text{ kJ/kg} + 2.5 * \left(645 \text{ kJ/kg} - 9 \text{ kJ/kg}\right) + 178 \text{ kJ/kg} - 915 \text{ kJ/kg}$$

$$q_{do} = 1548 \text{ kJ/kg}$$

The evaporator and condenser heat amount to:

$$q_e = h_{1'} - h_2 = 1311 \text{ kJ/kg} - 194 \text{ kJ/kg} = 1117 \text{ kJ/kg}$$

$$q_c = h_{7'} - h_2 = 1340 \text{ kJ/kg} - 194 \text{ kJ/kg} = 1146 \text{ kJ/kg}$$

The energy balance yields:

$$q_{do} + q_e + w_p - q_c - q_{ao} = (1548 + 1117 + 3 - 1146 - 1521) \text{ kJ/kg} = 1 \text{ kJ/kg}$$

The COP amounts to:

$$\text{COP} = q_e / q_{do} = 0.72$$

This value represents an increase of 26% compared to the single-stage absorption heat pump of Example 4.2. At the same time, the highest temperature in the desorber-absorber heat exchange unit is higher, too. The higher COP is obtained by using a higher availability driving heat in the desorber. This measure of increased temperature for the driving heat allows, in a single-stage system, the rise of the absorber and/or condenser temperatures (i.e., the temperature lift is increased). If this is not desired, desorber-absorber heat exchange can be employed and the COP increased instead.

It was assumed that all the heat released in the absorber above the temperature T_9 can actually be utilized in the desorber. That is usually the case for ammonia/water systems. However, if the desorber heat requirement in the overlapping temperature range is smaller than the amount of heat the absorber can reject, not all of q_g can be used for internal heat exchange. To confirm whether or not the internal heat exchange is limited by the desorber or absorber, the amount of heat the desorber can accept in the overlapping temperature range has to be calculated, too. The procedure is quite analogous to the one described for q_g.

Heat Exchange between Weak and Strong Absorbent by Desorption and Absorption.
So far, internal heat exchange was based on heat conduction only. However, a very effective heat transfer can be obtained by transferring latent heat by evaporation and condensation. This heat transfer process is accompanied by mass transfer. It is the same process that occurs in a heat pipe. This requires a transfer of mass as well. The heat exchange between rich and poor solution can be accomplished in a similar manner by desorbing working fluid from the hot solution and absorbing it in the cold solution.

4.6 LIMITATIONS OF SINGLE-STAGE ABSORPTION HEAT PUMPS

The operating parameters of temperature, pressure, and concentration within the various exchange units are closely coupled for the following reason. In single-stage absorption systems, the amount of availability provided by one exchange pair has to be processed in the second pair of exchange units using identical working-fluid flow rates. For a resorption unit as shown in Figure 4.11 with $T_1 = T'_1$, only two temperatures and one pressure or two pressures and one temperature can be selected independently from the set of three temperatures and two pressures. The two remaining variables and the average compositions ξ_1 and ξ_2 are determined by the solution field. If $\xi_1 = 1$, only two temperatures or two pressures or one temperature and the unrelated pressure can be selected independently. If absorber and resorber temperatures are not identical, there are four degrees of freedom among the eight variables (four temperatures, two pressures, two compositions). For $\xi_1 = 1.0$, there are three degrees of freedom among the seven variables.

With the help of the solution fields of NH_3/H_2O, in Figure 2.7, Chapter Two, it can be realized that a single-stage absorption heat pump can produce cooling to a lowest temperature of $-25°C$ (P = 1.6 bar), for a driving heat of $100°C$ and heat rejection at $30°C$ (P = 12 bar). Lower refrigeration temperatures can be obtained by two stage systems, as shown in Chapter Seven. Further, for a generator heat input at $180°C$, heat rejection at $35°C$, and an evaporator temperature of $10°C$, cooling efficiency of 3.7 can theoretically be obtained, whereas, a single-stage unit cannot achieve efficiencies higher than about 0.6. As can be seen from the solution field for NH_3/H_2O, the absorber heat is available at such a high temperature level that it can be used to drive a second stage as shown in Chapter Seven. The system $H_2O/Li Br$, Figure 7.11, produces, for similar evaporator temperatures and a generator temperature of $150°C$, a condenser temperature of $95°C$ sufficient to power a second stage which results in a higher COP, Figures 7.11 and 7.12, Chapter Seven. Means to extend the limits of single-stage absorption systems can be found in Chapters Six and Seven.

4.7 COMPARISON OF THE PERFORMANCE OF VARIOUS SYSTEMS

A. Vapor-Compression Heat Pump

In order to state a realistic COP, based on the consumption of primary energy, the efficiency of the power plant η_{pp} has to be taken into consideration. The COP of the heat pump is here η_{wp}:

$$\xi = \eta_{wp}\eta_{pp} = 3 \cdot 0.33 = 1$$

The heat pump absorbs, at its location, approximately as much heat from the surroundings as the power plant rejects as waste heat to the surroundings. An increase in efficiency can be achieved when the "power plant" is installed at the location of the heat pump (see B, below). Although electric vapor-compression heat pumps do not provide savings in primary energy, they are of benefit, since nuclear or coal-fired power plants or solid-waste combustion plants rarely can be installed close to the consumer.

B. Vapor-Compression Heat Pump with Internal Combustion Engine and Use of the Waste Heat

A diesel engine converts 100% primary energy into the following other forms of energy:

37% power

26% heat in cooling water

19% useful heat of the exhaust

10% exhaust losses

8% radiation losses

When the vapor-compression heat pump, powered by a diesel engine, shows an efficiency of 3, the overall efficiency of the complete system, including the heat in the cooling water and the exhaust, will be

$$COP = 3.0 \quad 0.37 + 0.26 + 0.19 = 1.56$$

C. Absorption Heat Pump

A realistic upper limit for a single-stage absorption heat pump will be

$$COP = 2 \quad 0.75 = 1.5$$

The factor 0.75 accounts for the following losses:

1. Irreversibilities within the unit such as pressure drops, finite pressure differences for absorption and desorption, incomplete internal heat exchange, etc.

2. Characteristic properties of the solution field, for example, the heat of solution which causes the vapor pressure curves to be steeper for absorber and generator than for the pure working fluid

Considering further a burner efficiency of 0.8 to account for exhaust and radiation losses, the overall efficiency will decrease to 1.2. This can be increased to 1.3 by using a flue gas heat exchanger. A further increase in the efficiency of absorption heat pumps can be achieved by multistage systems which will be discussed in detail in the following chapter. The generator-absorber heat exchange cycle (GAX) belongs to the multistage systems, Chapter Seven.

D. Vapor-Compression and Absorption Heat Pump with Internal Combustion Engine

When, as in case B, the waste heat is used to drive an absorption heat pump, the overall efficiency increases in the following way:

$$COP - 3 \quad 0.37 + 1.5 \quad (0.26 + 0.19) = 1.78$$

As will be shown later, both heat pumps can be integrated within one unit.

4.8 COP EVALUATION WITH LIMITED DATA

Heat ratios of absorber heat pumps and heat transformers are generally calculated by enthalpy balances using enthalpy-concentration charts. Only the First Law is used explicitly, but the COPs never violate the Second Law requirements. The Second Law is implicitly taken into account by using actual enthalpy values of real fluids. The COP is determined as the ratio of differences of large numbers, the enthalpy values. Therefore, the enthalpy values require a higher accuracy than that expected for the COP. The Second Law is independent of fluid properties. But, inherently, it is recalculated for every case requiring an extensive amount of experimental data.

In the following, the Second Law is explicitly used from the very beginning. The accuracy required for the fluid data and the amount of data is considerably reduced compared to the conventional approach. For new fluid pairs or mixtures, rather fast estimates for the COP are possible.

4.8.1 The First and Second Laws for the Absorption Heat Pump

Figure 4.33 shows temperatures and pressures for the four exchange units of an absorption heat pump. The First and Second Laws can be written as shown in Equation 4.9.
First Law:

$$q_2 - q_1 - \hat{q}_1 + q_0 - q_r + w = 0 \tag{4.9}$$

Second Law:

$$-\frac{q_2}{t_2} + \frac{q_1}{t_1} + \frac{\hat{q}_1}{\hat{t}_1} - \frac{q_0}{t_0} + \frac{q_r}{t_r} = \sum_1^n \delta s_i \tag{4.10}$$

with q_i, $\delta s_i \geq 0$ (for q_i, see Figure 4.33)

w = pumping power

q_r = heat of rectification

t_r = external temperature at which heat of rectification is removed

$\Sigma \delta s_i$ = sum over all entropy producing process steps including heat transfer to the external heat carrier fluids

The quantities q_i, w, and δs_i are actually rates and are based on one unit of mass of circulating working fluid.

The temperatures t_i are exactly defined averages over the temperature spread $t_i^2 - t_i^1$ of the heat carrier fluids:

$$\frac{1}{t_i} = \frac{1}{q_i} \int_{t_i^1}^{t_i^2} \frac{c_i dt}{t} \qquad q_i = \int_{t_i^1}^{t_i^2} c_i dt \tag{4.11}$$

For most cases, the average temperature is a good approximation with an accuracy better than 1%:

$$t_i \approx \frac{1}{2}\left(t_i^1 + t_i^2\right) \tag{4.12}$$

In the following, the pumping power w and the heat of rectification q_r will be neglected. This is an acceptable approximation for the working fluid pair H_2O/LiBr. Furthermore, it will be assumed that: $t_1 = \hat{t}_1$ and $T_1 = \hat{T}_1$.

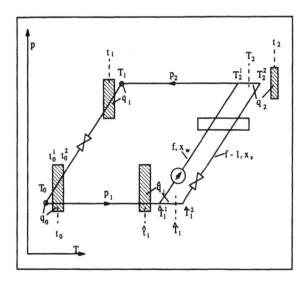

Figure 4.33 Temperatures and pressures for an absorption heat pump.

The following entropy producing steps can be identified:

δs_1 = desuperheating of the fluid between T_2 and T_1

δs_2 = cooling of the fluid from T_1 to T_0 (throttling)

δs_3 = heating of the fluid from T_0 to T_1

δs_4 = heating or cooling of the weak (in respect to absorbent) solution to T_2^1

δs_5 = cooling or heating of the strong (in respect to absorbent) solution to T_1^2

δs_6 = entropy generation in the solution heat exchanger

δs_7, δs_8, δs_9, δs_{10} = heat transfer between heat carrier fluids and working fluids

δs_{11} = all others, for example, irreversibilities of the pump, pressure drops, heat conduction, etc. (δs_{11} will be neglected in the following)

Eliminating $q_1 + \hat{q}_1$ from Equation 4.9 and 4.10 yields the following equation for the heat q_2 and the efficiency $\eta = q_0/q_1$ for **refrigeration**:

$$q_2 = q_0 \frac{t_1 - t_0}{t_0} \bigg/ \frac{t_2 - t_1}{t_2} + \frac{t_1 t_2}{t_2 - t_1} \sum_1^{10} \delta s_i \tag{4.13}$$

$$\eta = \frac{\eta^r(t)}{\left(1 + \frac{t_0 t_1}{t_1 - t_0} \sum_1^{10} \delta s_i / q_0 \right)} \tag{4.14}$$

with the efficiency $\eta^r(t)$ for a reversible machine, operating between t_2, t_1 and t_0:

$$\eta^r(t) = \frac{\dfrac{t_2 - t_1}{t_2}}{\dfrac{t_1 - t_0}{t_0}} \tag{4.15}$$

The quality of the process (effectiveness), including heat transfer, can be measured by the following ratio:

$$g(t) = \eta/\eta^r(t) = \frac{1}{\left(1 + \frac{t_0 t_1}{t_1 - t_0} \sum_1^{10} \delta s_i / q_0\right)} \tag{4.16}$$

The entropy generated by the heat transfer is the sum of four terms:

$$\sum_7^{10} \delta s_i = q_2\left(\frac{1}{T_2} - \frac{1}{t_2}\right) + q_1\left(\frac{1}{t_1} - \frac{1}{T_1}\right) + \hat{q}_1\left(\frac{1}{\hat{t}_1} - \frac{1}{\hat{T}_1}\right) + q_0\left(\frac{1}{T_0} - \frac{1}{t_0}\right) \tag{4.17}$$

This is an exact relationship if, for the internal temperatures $\hat{T}_1$ and T_2, the entropic averages are used, as shown in Equation 4.11. Yet, the heat distribution over internal temperature T must be inserted. For most cases, the average temperatures are good approximations, even if heat production is not completely homogeneous over the temperature interval; for example, over $\hat{T}_1^2 - T_1^1$, inserting Equation 4.17 into Equation 4.10 and eliminating $q_1 + \hat{q}_1$, now the following set of equations holds:

$$q_2 = q_0 \frac{\frac{T_1 - T_0}{T_0}}{\frac{T_2 - T_1}{T_2}} + \frac{T_1 T_2}{T_2 - T_1} T \sum_1^6 \delta s_i \tag{4.18}$$

$$\eta = \frac{\eta^r(T)}{\left(1 + \frac{T_0 T_1}{T_1 - T_0} \frac{\sum_1^6 \delta s_i}{q_0}\right)} \tag{4.19}$$

with the efficiency $\eta^r(T)$ of a reversible machine, operating between the internal temperature levels T_2, T_1, and T_0:

$$\eta^r(T) = \frac{\frac{T_2 - T_1}{T_2}}{\frac{T_1 - T_0}{T_0}} = \frac{\left(\frac{1}{T_1} - \frac{1}{T_2}\right)}{\left(\frac{1}{T_0} - \frac{1}{T_1}\right)} \tag{4.20}$$

The value of η from Equation 4.19 is identical to that of Equation 4.14.

Only two of the three temperatures T_0, T_1, and T_2 can be chosen independently; the third is determined by the individual properties of the solution field. Equation 4.20 can be expressed by the slopes of the isosteres for the pure fluid and the average value of the slopes for the strong and weak solution $\bar{r} + 1$.

$$\eta^r(T) = \frac{\bar{r}}{\bar{r} + 1} = \frac{1}{1 + \frac{1}{\bar{r}}} \tag{4.21}$$

Far from the critical point is the slope identical with the heat of vaporization r. This is the case for H_2O around room temperature.

Equation 4.18 and 4.13 can be interpreted as follows: The first term represents the minimum amount of driving heat, q_2, required for a given refrigeration power, q_0. Due to irreversibilities, the required heat input is increased proportionally to the entropy production. The proportionality factor $T_1T_2/(T_2 - T_1)$ differs from that found in calculating exergy losses. From Equation 4.19 a quality factor (an effectiveness) for the process without accounting for heat transfer can be defined as:

$$g(T) = \frac{\eta}{\eta^r(T)} = \frac{1}{\left(1 + \frac{T_0 T_1}{T_1 - T_0} \frac{\sum_1^6 \delta s_i}{q_0}\right)} \tag{4.22}$$

No ambient temperature t_u appears in this equation, as it does in an exergy analysis.

The entropy increase δs_1 is determined by the following integral:

$$\delta s_1 = \frac{1}{T_1} \int_{T_1}^{T_2} c_p dT - \int_{T_1}^{T_2} \frac{c_p dT}{T} \tag{4.23}$$

Taking the average for c_p and expanding to the lowest order in $T_2 - T_1$, the following relationship is obtained:

$$\delta s_1 = c_p \left(\frac{T_2 - T_1}{T_1} - \ln \frac{T_2}{T_1}\right) = \frac{c_p}{2} \frac{(T_2 - T_1)^2}{T_1 T_2} + \ldots \tag{4.24}$$

Similarly, δs_2 to δs_6 can be calculated:

$$\delta s_2 = \frac{c}{2} \frac{(T_1 - T_0)^2}{T_0 T_1} + \ldots \tag{4.25}$$

$$\delta s_3 = \frac{c_p}{2} \frac{(\hat{T}_1 - T_0)^2}{T_0 \hat{T}_1} + \ldots \tag{4.26}$$

where c and c_p are the specific heat capacity of the liquid and gaseous working fluid, respectively.

For the total entropy created in the solution circuit, the following expression is obtained:

$$\sum_4^6 \delta s_i = \frac{1}{2} \left[fc_w + (f-1)(1-2\eta_s)c_s\right] \frac{(T_2 - \hat{T}_1)^2}{T_2 \hat{T}_1} + \eta_s(f-1)c_s \frac{(\hat{T}_1^2 - \hat{T}_1^1)^2}{\hat{T}_1 T_2} \tag{4.27}$$

where η_s = efficiency of the solution heat exchanger, f = circulation ratio, c_s, c_w = specific heat capacity of the strong and weak solution. The last term in Equation 4.27 is small, compared to the first one and may be neglected in most cases.

The refrigeration capacity per unit mass after throttling is given by:

$$q_0 = r\left(1 - \frac{c(T_1 - T_0)}{r}\right) \tag{4.28}$$

Inserting Equations 4.24 to 4.28 into Equations 4.18 and 4.19 yields the final equations for the total heat input q_2 and the efficiency of the refrigerator.

The denominator in Equation 4.19, which is a correction to the reversible (Carnot) efficiency, is a simple function of a set of nondimensional fluid parameters listed in Table 4.4. The first two parameters are characteristic of the working fluid only. For the third and forth parameters, the specific heat capacities of the absorber fluid must be known. With the fifth parameter, the Carnot efficiency can be calculated without specifying the three temperatures T_0, T_1, T_2 explicitly. Only the ratio of the slopes of vapor pressure lines (or the temperature differences $T_2 - T_1$ and $T_1 - T_0$ and one temperature) must be known with the same accuracy as required for the COP. Since all the parameters listed in Table 4.4 are corrections of the order of 10% or less (see Table 4.5), an accuracy of an order of magnitude lower than that for enthalpies is sufficient. If the pumping power must be considered, an additional fluid parameter is needed. The absolute pressures at which the machine operates are only relevant for this parameter.

For the Tables 4.4 to 4.7, the following input data have been used: Fluid pair: $H_2O/LiBr$; c = 4.19 kJ/kgK, C_p = 1.90 kJ/kgK, r = 2490 kJ/kg, T_0 = 2°C, T_1 = 38°C, T_2 = 80°C, x_s = 0.595, x_w = 0.570, c_s = 1.97 kJ/kgK, c_w = 2.09 kJ/kgK, f = 16, $\hat{T}_1^2 - \hat{T}_1^1$ = 6 K, η_s = 0, 8, $T_i - t_i$ = 5 K.

Table 4.5 shows that the additional heat input caused by the irreversibilities δs_1, δs_2, and δs_3 amounts only to about 5% of the total heat input. This holds for H_2O as working fluid and changes appreciable for NH_3 or R22. The largest term results from the solution circuit. Again, this term strongly increases for other pairs.

In Table 4.6, the reversible COPs and three realistic COPs for different heat exchanger efficiencies η_s are listed. The reversible value 0.91 is below 1.0 due to the increased slope of the vapor pressure lines with LiBr. The value 0.76 is typical for a heat exchanger efficiency of 0.8. Even for η_s = 0, the COP stays as high as 0.58; which is only the case for $H_2O/LiBr$. Without heat transfer, the effectiveness is rather high, i.e., the COP reaches about 80% of the Carnot value.

4.8.2 The COP and Effectiveness for the Absorption Heat Pump

Using the relationship

$$\eta_{HP} = 1 + \eta \tag{4.29}$$

the COP and a quality factor (effectiveness) of the absorber heat pump can easily be calculated as:

$$\eta_{HP} = \frac{\dfrac{T_2 - T_0}{T_2}\,1 + \dfrac{T_0 T_2}{T_2 - T_0}\sum_1^6 \delta s_i / q_0}{\dfrac{T_1 - T_0}{T_1}\,1 + \dfrac{T_0 T_1}{T_1 - T_0}\sum_1^6 \delta s_i / q_0} \tag{4.30}$$

with the COP for the completely reversible heat pump:

$$\eta_{HP}^r = \frac{\dfrac{T_2 - T_0}{T_2}}{\dfrac{T_1 - T_0}{T_1}} \tag{4.31}$$

and an effectiveness:

Table 4.4 Nondimensional Fluid Parameters
for H$_2$O/LiBr

1	2	3	4	5	6
cT_1/r	c_pT_1/r	$(f-1)c_sT_1/r$	$f\,c_wT_1/r$	$1/r$	pump parameter
0.52	0.24	3.74	4.18	0.10	0.0

Table 4.5 Heat Input q$_2$ per Refrigeration Capacity q$_0$
for a Refrigerator or Heat Pump

	q_2/q_0	$1/\eta^r(T)$	$\dfrac{T_1\,T_2}{T_2-T_2}\dfrac{\delta s_1}{q_0}$	$\dfrac{T_1\,T_2}{T_2-T_2}\dfrac{\delta s_2}{q_0}$	$\dfrac{T_1\,T_2}{T_2-T_2}\dfrac{\delta s_3}{q_0}$	$\dfrac{T_1\,T_2}{T_2-T_2}\dfrac{\delta s_4+\delta s_5+\delta s_6}{q_0}$
HP	1.32	1.10	0.017	0.036	0.016	0.152
	120%	100%	1.5%	3.3%	1.4%	13.8%
HT	0.90	1.10	0.016	0.033	0.015	0.139
	81.8%	100%	1.4%	3.0%	1.3%	12.5%

Note: (HP: Equation 4.18, $\eta_s = 0.8$) and heat output q$_2$ per heat output q$_0$ for a heat transformer

Table 4.6 Heat Ratios q$_0$/q$_2$ and Quality Factors g
(Effectiveness) for an Absorption Refrigerator
(Equations 4.14, 4.15, 4.16, 4.19, 4.20, 4.21)

	$\eta_r(T)$	$\eta_s = 0.8$	$\eta_s = 1.0$	$\eta_s = 0.0$	$\eta_T^r(t)$
$\eta = q_0/q_2$	0.91	0.76	0.83	0.53	1.56
$g(T)$	—	83%	91%	64%	—
$g(t)$	—	49%	53%	37%	—

Note: (HT: Equation 4.38, $\eta_s = 0.8$) $|T_i - t_i| = 5$ K).

Table 4.7 Heat Ratios q$_2$/(q$_1$ + q$_1$) and Quality Factors g
(Effectiveness) for a Heat Transformer

	$\eta_T^r(T)$	$\eta_s = 0.8$	$\eta_s = 1.0$	$\eta_s = 0.0$	$\eta_T^r(t)$
$\eta_T = \dfrac{q_2}{(q_1+\hat{q}_1)}$	52.5%	47.6%	50.1%	34.3%	60.0%
$g_T(T)$	—	90.7%	95.5%	65.3%	—
$g_T(t)$	—	79.1%	83.1%	11.1%	—

Note. (Equations 4.33 to 4.38, $|T_i - t_i| = 5$ K).

$$g_{HP}(T) = \eta_{HP}/\eta_{HP}^r = \frac{(1+\eta)}{(1+\eta^r(t))} \qquad (4.32)$$

Table 4.5 holds also for heat pumps. In Table 4.6, the first line must be increased by 1.0. The quality factors for a heat pump approaches 1 for a reversible machine, but not zero for $\eta \to 0$. This is caused by the definition of η_{AP} in Equation 4.29 which approaches 1 for the worst machine. It is, therefore, recommended that one use Equation 4.22 instead of Equation 4.32 to measure the Second Law performance of a machine.

4.8.3 First and Second Laws for the Heat Transformer
According to Figure 4.34, the First and Second Laws of an absorption heat transformer are identical to Equation 4.9 and 4.10 (neglecting q$_r$ and w), except that the sign for the heat fluxes q$_i$ is reversed:

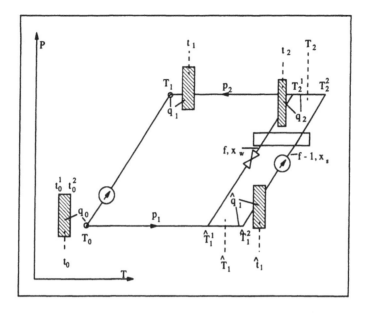

Figure 4.34 Temperatures and pressure for a heat transformer.

First Law : $\quad -q_2 + q_1 + \hat{q}_1 - q_0 = 0$ (4.33)

Second Law : $\quad \dfrac{q_2}{t_2} - \dfrac{q_1}{t_1} - \dfrac{\hat{q}_1}{\hat{t}_1} + \dfrac{q_0}{t_0} = \displaystyle\sum_1^n \delta s_i \quad$ with $q_i, \delta s_i \geq 0$ (4.34)

The entropy production due to heat transfer is now given by:

$$\sum_7^{10} \delta s_i = q_2\left(\frac{1}{t_2} - \frac{1}{T_2}\right) + q_1\left(\frac{1}{T_1} - \frac{1}{t_1}\right) + \hat{q}_1\left(\frac{1}{\hat{T}_1} - \frac{1}{\hat{t}_1}\right) + q_0\left(\frac{1}{t_0} - \frac{1}{T_0}\right)$$ (4.35)

With the expansion to the lowest order in $T_2 - _1$ or $\hat{T}_1 - T_0$, respectively, the entropy production for the heat transformer δs_1 to δs_6 is identical to that of the heat pump. Thus, Equations 4.21 to 4.27 can be used in Equation 4.34.

The heat q_0 rejected at the condenser is, in contrast to Equation 4.28 larger than r, namely:

$$q_0 = r\left(1 + \frac{c_p\left(\hat{T}_1 - T_0\right)}{r}\right)$$ (4.36)

Equations 4.33 and 4.34 can now be solved for q_2 and $\eta_T = q_2/(q_1 + \hat{q}_1)$ by eliminating q_0. Yet, the results are more transparent and can be compared more easily with those of the heat pump if first the heat ratio q_2/q_0 is determined, as has been done for the refrigerator. The efficiency of the heat transformer can then be calculated with the equation:

$$\eta_T = \frac{q_2/q_0}{1 + q_2/q_0} = \frac{1}{1 + q_0/q_2}$$ (4.37)

Inserting Equation 4.35 into Equation 4.34 and eliminating $q_1 + q_1$ yields:

$$q_2 = q_0 \frac{\dfrac{T_1 - T_0}{T_0}}{\dfrac{T_2 - T_1}{T_2}} - \frac{T_1 T_2}{T_2 - T_1} \sum_1^6 \delta s_1 \tag{4.38}$$

Equation 4.38 is identical with Equation 4.18, except for the differing value of q_0 (Equation 4.36) and the minus sign of the last term. Clearly, for a heat transformer irreversibilities will reduce the heat output q_2.

Combining Equation 4.38 with Equation 4.37, one finds for the efficiency of the heat transformer:

$$\eta_T = \frac{\dfrac{T_1 - T_0}{T_1}}{\dfrac{T_2 - T_0}{T_2}} \frac{1 - \dfrac{T_0 T_1}{T_1 - T_0} \sum_1^6 \delta s_i / q_0}{1 - \dfrac{T_0 T_2}{T_2 - T_0} \sum_1^6 \delta s_i / q_0} \tag{4.39}$$

This equation is the reciprocal of the heat pump efficiency Equation 4.30, but with minus signs in front of the sum for the entropy increases.

The completely reversible heat transformer has the efficiency:

$$\eta_T^r = \frac{\dfrac{T_1 - T_0}{T_0}}{\dfrac{T_2 - T_0}{T_2}} = \frac{\dfrac{1}{T_0} - \dfrac{1}{T_1}}{\dfrac{1}{T_0} - \dfrac{1}{T_2}} \tag{4.40}$$

The quality or effectiveness of the process can be measured by the following ratio:

$$g_T(T) = \frac{\eta_T}{\eta_T^r} = \frac{1 - \dfrac{T_0 T_1}{T_1 - T_0} \sum_1^6 \delta s_i / q_0}{1 - \dfrac{T_0 T_2}{T_2 - T_0} \sum_1^6 \delta s_i / q_0} \tag{4.41}$$

The analogous equation holds for the effectiveness $g_T(t)$, including heat transfer replacing the internal temperatures T_i by the external temperatures t_i and including the heat transfer in the sum of the irreversibilities:

$$g_T^{(t)} = \frac{\eta_T}{\eta_T^r(t)} \tag{4.42}$$

The fluid pair by itself may be characterized by $g(T)$, yet taking for the heat exchanger efficiency η_s in Equation 4.27 the best ($\eta_s = 1$) or the worst ($\eta_s = 0$) value.

The reversible efficiency η_T^r can be expressed by the slopes of the isosteres in vapor pressure diagrams:

$$\eta_T = \frac{1}{1 + \dfrac{\bar{r}}{\bar{r} + 1}} = \frac{1 + \dfrac{1}{\bar{r}}}{2 + \dfrac{1}{\bar{r}}} \tag{4.43}$$

For the heat transformer, the same nondimensional parameters as in Table 4.4 can be used. In Table 4.5, the second line shows how much the heat output q_2, relative to the heat output

q_0 is *reduced* by irreversibilities. The small differences in the third to the sixth columns between heat pump (HP) and heat transformer (HT) result from different values of q_0, namely $q_0 (HP)/q_0 (HT) = 0.915$. For the heat transformer, the sum of columns 3 to 6 is subtracted from column 2 to yield column 1.

In Table 4.7, heat ratios of a heat transformer for three heat exchanger efficiencies and for reversible processes are listed together with quality factors.

The high values of $\eta_T = 47$ to 48%, that is about 90% of Carnot efficiency, have been reached in laboratory prototypes as well as in large commercial plants.

4.8.4 Conclusion

Starting from the First and Second Laws, simple equations for the heat ratios (COPs) of absorption refrigerators, absorption heat pumps, and absorption heat transformers have been derived. These equations are particularly useful if only a limited amount of fluid data or data with limited accuracy are available.

As a by-product, a Second Law analysis is presented, which does not have the problems that are commonly encountered for an exergy analysis. In such an analysis, exergy losses are calculated. The exergy is the work which could be gained from heat in a reversible power station working against the lower temperature level t_u. The exergy losses are proportional to the entropy production, with t_u as the proportionality factor. Therefore, the choice of t_u leaves an ambiguity. The effectiveness of a machine, defined as the real efficiency, in respect to that of a reversible machine, should not depend on an arbitrarily chosen or accidentally existing temperature t_u.

Of more relevance than the problem of how much potential work is lost in a process step are the following questions: How much additional heat at T_2 must be supplied due to irreversibilities? Or, how much is the heat output at T_2 reduced due to irreversibilities?

The answer can be found in Equations 4.18 or 4.38 of this chapter. These additional heat quantities are proportional to $\Sigma \delta s_i$, as in the exergy loss. Yet, if these heat quantities are converted into exergies, one finds that the factors by which the entropy increase must be multiplied can qualitatively and quantitatively differ widely from t_u. In the analysis of this chapter, only process temperatures enter into the equations for efficiency or effectiveness. The equations are independent from the ambient.

REFERENCES

1. This presentation is an extract of a more detailed paper by the author on: "The Heat Ratios of Absorber Heat Pumps and Heat Transformers from First and Second Law", submitted.

2. Recommended reading: Introduction to the author's paper "What Needs to Be Known about Working Fluids to Calculate COPs?", Proceedings of the 1987 International Energy Agency Heat Pump Conference. Orlando, FL, April 28–30, 1987, Zimmermann, K. H., Ed., Lewis Publishers, Chelsea, MI.

3. See, e.g., D.L. Hodgett, "New Working Pairs for Absorption Processes", Proceedings of a Workshop in Berlin, (1982), Editor: Swedish Council for Building Research.

4. P. Riesch et al., "Part Load Behavior of an Absorption Heat Transformer", presented at the Third Intern. Symposium on the Large Scale Application of Heat Pumps, Oxford, March 1987.

5. Data sheets by Sanyo, Hitachi Zosen, or GEA.

6. See, e.g., W. Fratscher and J. Beyer, "Stand und Tendenzen bei der Anwendung und Weiterentiwicklung des Exergiebegriffs", Chem. Technik, 33, pp. 1–10 (1981).

7. Niebergall, W. (1981): Sorptions Kältemaschin, Handbook der Kültetechnik Bd. 7 Springer Verlag, Berlin.

8. Bosnyaković, F. (1971): Technische Thermodynamik Teil 2. Steinkopff Verlag, Dresden.

9. Bosnyaković, F., K.F. Knoche (1988): Technische Thermodynamik Teil 1. Leipzig, 7. Auflage.

10. Threlkeld, J. L. (1970): Thermal Environmental Engineering, Second Edition, Prentice Hall Inc., Englewood Cliffs, New Jersey.

11. Bogart, M. (1981): Ammonia Absorption Refrigeration in Industrial Processes. Gulf Publishing Company, Houston, Texas.

12. Planck, R., Kuprianov, J. (1948): DieKlein Kältemaschine, Springer Verlag, Berlin.

13. Altenkirch, E., Absorptions Kältemaschine, VEB Verlag Technik, Berlin.

14. Alefeld, G. (1987b): What needs to be known about fluid pairs to determine heat ratios of absorber heat pumps and heat transformers, In: *Proceedings of the 1987 International Energy Agency Heat Pump Conference* (K. H. Zimmerman, Ed.), Chap. 26, pp. 375-387. Lewis Publishers, Chelsea, MI.

CHAPTER FIVE

Rules for the Design of Multistage, Absorption Heat Pumps, Heat Transformers, Vapor-Compression Heat Pumps, Heat Engines, and Cascades

It is desirable to know all cycles which fulfill the requirements of a given application. In this chapter, design rules are developed which, within a given set of components, allow the finding of all competing cycles. The formulation of the rules are based on First, Second, and Mass Conservation Laws. The rules are developed to such an extent that their application is reduced to a mere combinatorial task. Three methods will be introduced. The first starts from building elements like condensers, evaporators, absorbers, generators, heat exchangers, compressors, pumps, and throttles. In the second method, large units, namely basic cycles, are used as design elements. The third method uses group theory as a method for generating new cycles. The second method is the most practical one; it will be used throughout this chapter. The design rules are explicitly formulated for this method. In addition, these design rules provide a classification scheme and permit study of all modes of operation for a given design.

5.1 INTRODUCTION

Traditionally, multistage absorption machines were used for cooling and refrigeration purposes to achieve higher temperature lifts or improved efficiency. Absorption machines are often considered to be complicated, with hard-to-understand process schemes. Multistage vapor-compression systems are used predominantly for refrigeration purposes when high temperature lifts are required. Only a few examples exist in which absorption and compression technologies are combined.

Examples are booster compressors in absorption machines and vapor-compression systems with solution circuits. In the past, the search for new cycles was based on trial and error. In recent years, an intensive search for new cycles, can be noticed. To facilitate the search for new cycles, a more systematic approach is introduced.

It is necessary to define several terms to facilitate the understanding of the classification system. The term "main component" refers to "exchange units", these are heat exchangers in which a phase change of the working fluid occurs. "Main component" may also refer to compressors or expansion machines. All other components such as pumps, bypasses, solution heat exchangers, rectifiers, throttles, valves, and receivers are termed "secondary components." The term "secondary component" does not include an economical or technical assessment of this component.

The main components of absorption machines and heat engine cycles can, in principle be connected arbitrarily by tubing for working fluid or for solution in the same way that electronic components can be connected to circuits and networks. Certain rules must be observed to achieve a useful combination of components. Even if stable operation is obtained, care must be taken to avoid connections which cause undesired losses in performance. For example, a tube connecting the vapor phase of the generator with the absorber of an absorption unit represents a detrimental connection, as does a tube connecting the condenser outlet to the absorber.

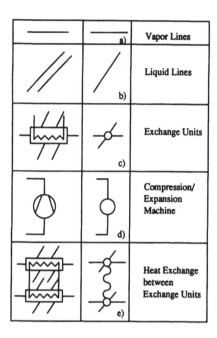

Figure 5.1 Examples of components used in heat conversion networks.

Finally, it should be possible to determine whether a given objective is achieved with the lowest first cost. This implies that all possible configurations which achieve the same result have to be known. Therefore, the combinational task has to include the principle of minimized entropy production.

The first column of Figure 5.1 shows the building blocks which can be combined to design heat conversion systems. The simpler symbols in the second column will be used later. Horizontal lines represent tubing carrying working fluid vapor. Diagonal lines represent tubing carrying liquid working fluid. Pairs of diagonal lines represent tubing carrying a mixture of working fluid and absorbent. For the sake of simplicity, such pairs often will also be represented by a single line. In Figure 5.1c, an exchange unit is fitted with connections for gaseous and liquid working or absorbent fluid. An expansion machine or compressor will be represented by a large circle (Figure 5.1d). Equipment providing heat exchange between two exchange units is indicated by a wavy line (Figure 5.1e).

All symbols are chosen so that the location of the exchange units in a diagram showing pressure on the ordinate and temperature on the abscissa indicates the relative pressure and relative temperature in that unit. The cycles shown in these diagrams can be seen as superimposed on solution fluids using the same coordinates. Therefore, the relative position of exchange units and their liquid line connections indicate also the relative concentration of the circulating solutions. Thus, the relative values of pressure, temperature, and composition prevailing in the various components are displayed. This statement does not hold for compressors and expansion machines. To avoid designs with built-in irreversibilities, the pressure drop is assumed initially to be negligible in the exchange units and in the horizontal connecting lines. Therefore, the lines carrying vapor have to be horizontal. The wavy lines indicating the internal heat exchange have to be vertical. (For detailed information on irreversibilities, see Section 5.3, below.)

Figure 5.2 shows an example of a configuration with a compression/expansion machine and six exchange units, two of which are connected by internal heat exchange. This configuration consists of two loops, having two independent working fluid streams (Kirchoff's Rule).

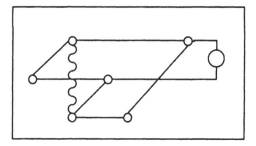

Figure 5.2 Example of a configuration with six exchange units, one compression/expansion machine, and internal heat exchange.

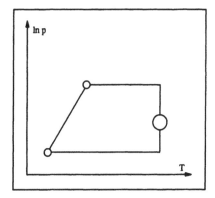

Figure 5.3 Graph for vapor-compression heat pump or heat engine.

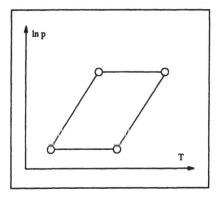

Figure 5.4 Graph for absorption heat pump or heat transformer.

All possible combinations of building blocks from Figure 5.1 can be constructed and compared. This allows the finding of all possible configurations containing a given number of exchange units and/or compression/expansion machines. This is the first method.

However, there is a faster design method. Instead of starting from building blocks, as seen in Figure 5.1, new configurations are constructed by combining basic cycles according to Figures 3.17 (Chapter Three) and 4.3 (Chapter Four) and which are shown in Figures 5.3 and 5.4 with the symbols defined in Figure 5.1. This second method is a short cut; therefore, care has to be exercised to ensure that really all possible combinations are found. For this purpose, the design rules are formulated.

5.2 DESIGN RULES FOR THE CONFIGURATION OF HEAT CONVERSION SYSTEMS

The rules are introduced first and are substantiated by examples later in this chapter.

Definitions:

1. A "complete cycle" is a configuration of exchange units and/or compression/expansion machines which allows a steady-state operation which theoretically can be a reversible operation. In the simplest possible case, a complete cycle is a single stage absorption cycle, Figure 4.3, or a single-stage compression/expansion cycle, Figure 3.17. Both cycles will be termed "basic cycles". In the following rules , the term "complete cycle" may be replaced by the more specialized term "basic cycle".

2. The number of stages of a certain configuration is given by the number of basic cycles needed to compose the configuration. (See comments on Rule 2, below, for details concerning the uniqueness.)

3. The elements of a class of cycles are determined by the identical topology of the connections for working fluid streams between the exchange units and the compression/expansion machines. (The term "identical topology" is explained in Section 5.3.2, below.)

4. A "cascade configuration" is an assembly of cycles connected by heat exchange only. In cascade configurations, the working fluids and the operating parameters of the various cycles can be selected independently. Multistage cycles are an assembly of cycles connected by heat exchange as well as by fluid streams. This implies that a certain exchange unit belongs to at least two complete cycles.

Rule 1
Any combination of exchange units and compression/expansion machines must be decomposable into a set of complete cycles. (Second Law: maximum possible conservation of availability.) An exchange unit or a compression/expansion machine can belong to more than one complete cycle (superposition). A given configuration can be extended by adding a basic or a more complicated complete cycle. At least one exchange unit or compression/expansion machine of this cycle must be connected heat-wise or fluid- and heat-wise to the original configuration. Any exchange unit of the first configuration can be coupled with any exchange unit of the second configuration.

Rule 2
Within a complete cycle, the direction and flow-rate of the working fluid stream can be chosen arbitrarily. The maximum possible number of working fluid streams can be found by applying Kirchhoff's Rule. When the configuration is composed of basic cycles only, the number of independent working fluid streams is the same as the number of the basic cycles. This statement is unique if a new basic cycle is coupled so that only one new loop for the working fluid is created. The number of independent working fluid streams determines the number of independent complete cycles and vice versa. The number of independent cycles is equal to or smaller than the number of complete cycles into which the configuration can be decomposed.

Rule 3a
An exchange unit can be removed from the configuration when the working fluid streams of the independent cycles to which the exchange unit belongs are selected, such that the phase

change occurring in this exchange unit of the stream of one cycle is canceled by the opposite phase change of the working fluid stream of a second cycle. For example, a certain exchange unit may operate in one cycle as an evaporator and in another as a condenser. These functions compensate each other if the flow rates are equal and the flow directions are opposite to each other. The same holds for an exchange unit that operates as an absorber or desorber, respectively, or for a machine that operates in one cycle as an expansion and in the other cycle as a compression machine. Rule 3a is not applicable for cascades.

Rule 3b

By a specific selection of both the ratios of working fluid flow rates and of the directions of the streams, partial or complete heat exchange between exchange units can be accomplished. For complete internal heat exchange, the number of temperature levels for heat exchange to or from the unit is reduced. Internal heat exchange leads to either increased efficiencies or higher temperature lifts.

Rule 4

All classes of two-stage machines can be created by connecting two basic cycles so that the two cycles have either one or two exchange units in common. All three-stage machines are obtained by combining the two-stage cycles with an additional basic cycle. The combined cycles have one, two, or three exchange units in common. All $(s + 1)$-stage units can be created by adding one basic cycle to an s-stage configuration so that there are anywhere from one to four $(s > 2)$ exchange units in common.

Using the same algorithm, all cascade configurations can be created. In this case, the connecting exchange units are only coupled by heat exchange. Rule 3a is therefore not applicable. In Appendix One found at the end of this text, options for internal heat exchange on one or two temperature levels between two basic cycles are listed. These options are applicable to two-stage cycles as well as to two-stage cascades. Cascades can be created by combining single as well as multistage configurations.

Comments:

1. Complete cycles may be arranged so that connections between exchange units pass through other exchange units without these units being counted. Technically, this may sometimes be desirable to achieve special effects such as preheating, precooling, or rectification of a gas stream by conducting this stream through an exchange unit belonging to another independent cycle. If this leads to an undesirable pressure drop, the gas stream can bypass the exchange unit.

2. The general form of the equation giving the number of independent working fluid streams reads: $z = 1 + x - y$ (x = number of connections between knots, y = number of knots). A knot is an exchange unit or a branch in a fluid stream. If a solution circuit is represented by one connection, then the net working fluid flow rate is determined as the difference of working fluid flow rates in the weak and strong absorbents. If a solution circuit is represented by two connections, then the solution circuit constitutes an additional independent stream. When the pressure drops are small, then bypasses to circumvent exchange units do not generate new modes of operation. Therefore, exchange units on the same pressure level can be combined into one common knot. This results in the following version of Kirchhoff's Law: $n = 1 + x - p$ (n = number of independent working fluids streams, x = number of connections between different pressure levels, p = number of pressure levels). For the application of this equation, a complete solution circuit is counted as one connection.

3. In order to apply Rule 3a, the net gaseous working fluid stream must be zero. This requirement is only approximately the same as the requirement that no net heat exchange takes place within the particular exchange unit since the circulating absorbent (if any) must be considered as well. To apply Rule 3b for complete internal heat exchange, the net flow rates of working fluid streams have to be determined from the requirement that exactly the amount of heat delivered by one exchange unit be completely absorbed by the other exchange unit. The internal heat exchange (whether it is complete or not) can only take place between exchange units which originally belonged to two different complete cycles. Unless Rule 3a is applied, some of the possible configurations which can be built from building blocks according to Figure 5.1 would be missed by applying Rule 1 only. Rule 3a is necessary, since two basic cycles are used as elementary building blocks instead of the building blocks shown in Figure 5.1.

5.3 GRAPHIC REPRESENTATIONS

5.3.1 Reversible and Irreversible Graphs

Initially, only main components will be represented in graphs used to design multistage systems. Secondary components can be added later. Figures 5.3 and 5.4 show graphs for basic compression/expansion cycles, and basic absorption cycles respectively (compare Figures 3.17 and 4.3). The small circles indicate exchange units, and the large circles represent the compression/expansion machines. The graphs display the process scheme and also the relative pressures, temperatures, and composition of the fluid mixture. One may envision the vapor pressure lines for the working fluid/absorbent combinations being superimposed. A horizontal shift of a pair of exchange units (Chapter Three, Section 6) means a change of the concentrations. Horizontal connections carry working fluid vapor, while diagonal connections carry liquid working fluid possibly dissolved in absorbent. The location of the compression/expansion machines is only indicative of the pressure levels, but not of the temperatures. For this chapter, the convention is used that the compression/expansion machine is always located towards the right (higher temperatures) of the exchange unit supplying or receiving the vapor. A power-producing process is therefore represented by a clockwise circulation of the working fluid, while a process requiring power input is represented by a counterclockwise circulation. If exchange units are connected with horizontal lines, it is indicated that no pressure drop has occurred. The same holds for connections between compression/expansion machines and exchange units.

The reversible graphs of Figure 5.3 and 5.4 will be used as building blocks for new configurations. A configuration may become distorted by irreversibilities such as pressure drops, temperature differences, and composition gradients. These processes do not generate new configurations. Figure 5.5a to d shows the basic graphs reflecting large pressure drops. The graphs in Figure 5.6 are excluded by the Second Law. Figure 5.7 shows distortions of reversible basic graphs of Figure 5.4 that are in compliance or contradiction to the Second Law. Figure 5.8 (double-effect machine) contains a graph, ABCD, which seems to be in contradiction to the Second Law, according to Figure 5.6. However, Figure 5.8 can be understood as a combination of the valid, but irreversible, graphs ABFE and CDFE.

Figure 5.9 presents graphs in which irreversibilities are caused by improper connections. Graphs of Figure 5.9e and f show an additional connection (possibly with flow control valve) between high and low pressure parts. If working fluid is transferred through this connection, the efficiency as compared to a reversible unit is reduced. However, these bypasses can be advantageous, for example, when a heat pump is switched to direct heating on very cold days. The vapor generated in the generator acts as a heat transfer medium and flows directly to the absorber to produce heat.

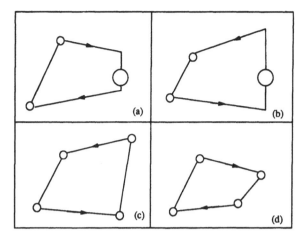

Figure 5.5 Irreversible graphs (pressure drop).

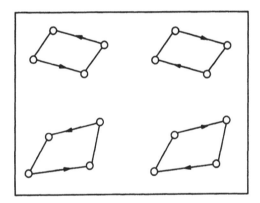

Figure 5.6 Irreversible graphs (Second Law).

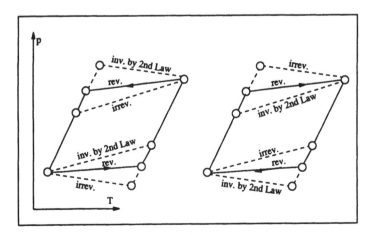

Figure 5.7 Distortion of reversible graphs.

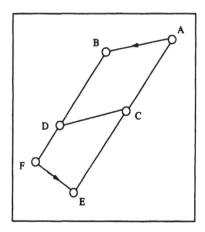

Figure 5.8   Distortion of the graph of a double-effect absorption system.

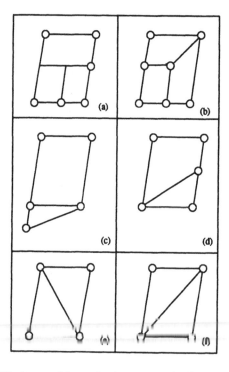

Figure 5.9   Irreversible graphs (counter-productive connections).

A further cause for irreversibility besides pressure drop is the exchange of heat between components, which are on different temperatures. Since working fluid and absorbent have finite heat capacities, there will always be heat transfer from a higher to a lower temperature level despite solution heat exchangers or internal heat transfer. This heat transfer is independent of the direction of the streams. In the graphs of Figures 5.10 and 5.11, this irreversible heat exchange is symbolized by wavy arrows. This heat transfer reduces the efficiency of a configuration.

In the following chapters, all possible configurations will be generated by combining reversible graphs. Pressure drop and heat exchange can be indicated later by distortions or wavy lines symbolizing internal heat transfer.

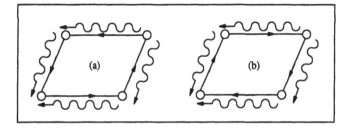

Figure 5.10   Irreversibility introduced by heat flow.

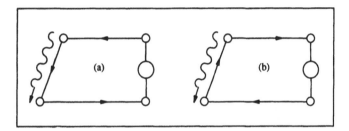

Figure 5.11   Graph with compression/expansion machine and irreversibility by heat flow.

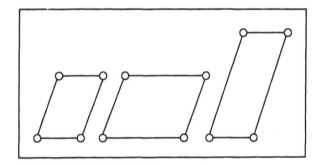

Figure 5.12   Topologically equivalent single-stage configurations.

5.3.2 Topological Equivalence

The topology of a graph is determined by the connections between exchange units and by the relative pressures and concentrations in the exchange units. Figure 5.12 shows topologically equivalent absorption cycles. Figure 5.13 shows two-stage cycles which are considered to be not topologically equivalent to those in Figure 5.14. Yet those in each figure are considered to topologically equivalent to each other. For topological equivalent configurations, horizontal and diagonal connectors can be shifted parallel as long as no other connection is crossed. The composition and pressure sequence $\xi_1 < \xi_2 < ...\xi <_n$, $P_1 < P_2 < ...<P_n$, assigned to the exchange units and the connection between them does not change during shifts within the P,T,ξ-field. On the other hand, the relative location with respect to temperature can be changed within certain limits for topologically equivalent cycles. This results in several versions of topologically equivalent configurations having different internal heat exchange paths.

The configurations shown side by side in Figures 5.12 and 5.14 are topologically equivalent.

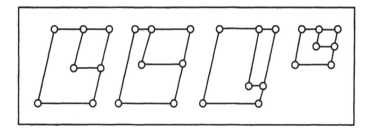

Figure 5.13 Topologically equivalent two-stage configurations.

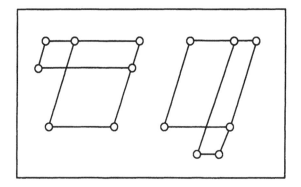

Figure 5.14 Topologically equivalent configurations (which are not topologically equivalent to those in Figure 5.13).

REFERENCES

1. Alefeld, G. (1982a): Regeln für den Entwurf von mehrstufigen Absorptionswmepumpen. In: Brennstoff-Wärme-Kraft, Bd.34, Nr. 2, S. 64-73.

2. Alefeld, F. (1983): Wärmeumwandlungssysteme, Skriptum zur Vorlesung am Institut für Technische Physik E 19, Physik-Department der Technischen Universität München.

3. Alefeld, G. (1982b): Kompressions-und Expansionsmaschinen in Verbindung mit Absorberkreisläufen. In: Brennstoff-Wärme-Kraft, Bd. 34, Nr. 3, S. 142.

4. Alefeld G. (1981): Der Wärmepumpentransformation, Brennstoff-Wärme-Kraft, vol. 33, pg. 486-490.

CHAPTER SIX

Cascade Configurations

Cascades in refrigeration applications are employed to pump heat across large temperature lifts. Typically, one refrigeration cycle pumps heat from the low temperature evaporator and rejects its waste heat on some intermediate temperature level which is below the temperature of the surroundings. A second refrigeration system picks up the waste heat of the first and lifts it to a temperature level high enough for rejection to the surroundings (Figure 6.1).

Figure 6.2 illustrates a cascade configuration of two power generation cycles. The high temperature cycle works with mercury as the working fluid, the low temperature cycle uses water. The heat of condensation of the mercury Rankine Cycle is supplied as boiler heat to the water Rankine Cycle. Power plants using mercury and water were successfully employed in the 1920s, but were abandoned for safety reasons. Recent investigations consider substituting potassium for mercury as a working fluid in the high temperature cycle.

In Figure 6.3, the absorption cycle is a heat transformer. The useful heat produced by the transformer is lifted to an even higher temperature level by a vapor-compression heat pump.

Figure 6.4 displays a cascade in which the heat rejected by a back-pressure Rankine Cycle operating with water as the working fluid is supplied as driving heat to an absorption refrigeration cycle. The data given for this cycle are typical for an air-conditioning system and may be used in cogeneration systems. In the application at Kennedy Airport in New York, the topping cycle is a gas turbine cycle.

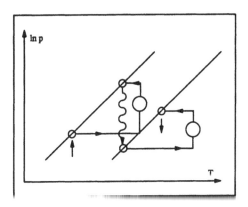

Figure 6.1 Cascade of two refrigeration systems.

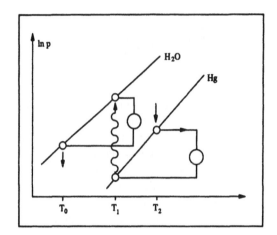

Figure 6.2   Cascade of two power generation cycles. The first cycle operates with mercury, the second with water.

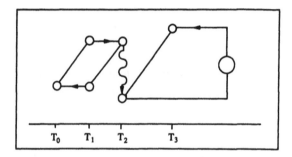

Figure 6.3   Cascade of an absorption cycle and a vapor-compression cycle.

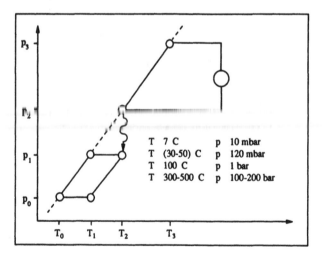

Figure 6.4   Cascade of a power plant and an absorption cycle (cogeneration of power and cooling).

REFERENCES

1. Threlkeld J. L., Thermal Environmental Engineering, Second Edition, Prentice Hall Inc., Englewood Cliffs, New Jersey, 1970.

2. Niebergall, W., Sorptions Kältemaschine Handbuch der Kültetechmik Bd 7 Springer Verlag, Berlin, 1981.

CHAPTER SEVEN

Two-Stage Absorption Heat Pumps and Heat Transformers

Two-stage absorption heat pumps and heat transformers have the potential to achieve higher temperature lifts or higher COPs than single-stage systems. The limits due to properties of the solution field often require multistage systems to meet the demands of a given application. In this chapter, all possible two-stage configurations based on the basic cycles in Figures 5.3 and 5.4 (Chapter Five) are introduced. In particular, their performance characteristics and operating modes are discussed and selected cycles are analyzed in detail.

7.1 CLASSIFICATION OF TWO-STAGE ABSORPTION UNITS

Figure 7.1 displays 26 topologically different configurations of two-stage absorption units. The configurations 7.1a to r can be obtained by coupling each exchange unit of a first basic cycle A with each exchange unit of a second basic cycle B (Figure 7.2). The exchange units of the basic cycles are numbered counterclockwise starting with the exchange unit of highest pressure and of highest temperature (Figure 7.2). The symbol A_1*B_1 means that the two exchange units numbered 1 are coupled by heat transfer when cascades are used or merged into one unit when multistage configurations are used. This coupling leads to two classes or two sets of topologically different configurations, Figures 7.1a,b. After coupling, the two cycles A and B are considered to be equivalent. Consequently, the operation A_i*B_j is identical to A_j*B_i. The result is that only 10 out of 16 options need to be considered. Of these ten cycles, eight have two topologically nonequivalent configurations, while the remaining two produce only one configuration each. Therefore, 18 topologically different configurations with the following characteristics can be constructed:

α.   18 classes (Figure 7.1a to r):

      7 exchange units

      2 working fluid streams

      3 pressure levels

      7 temperature levels (which may be reduced for some classes to three by applying Rule 3b).

If two exchange units of the basic cycles are coupled, the configurations shown in Figure 7.1s and t are obtained. They exhibit the following characteristics:

β.   1 class (Figure 7.1s):

      6 exchange units

      2 working fluid streams

      3 pressure levels

      6 temperature levels (which may be reduced to three by applying Rule 3b),

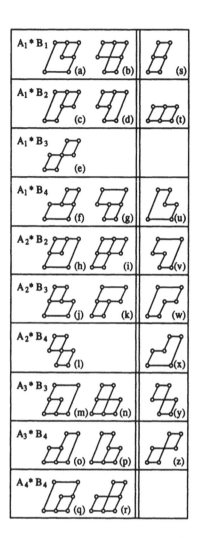

Figure 7.1 Two-stage absorption units.

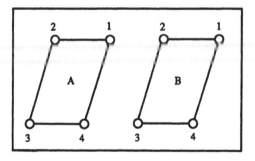

Figure 7.2 Basic cycles A and B.

γ.   1 class (Figure 7.1t):

6 exchange units

2 working fluid streams

2 pressure levels

6 temperature levels (which may be reduced to three by applying Rule 3b)

The class shown in Figure 7.1t is the only two-stage absorption heat pump with two pressure levels.

If Rule 3a is applied to the 18 classes of Figure 7.1a to r, sets of three classes result in one new class each. These exhibit the following properties:

δ.   6 classes (Figure 7.1u through z):

   6 exchange units

   1 working fluid stream

   3 pressure levels

   6 temperature levels (which may be reduced to three by applying Rule 3b).

Consequently, 26 classes of two-stage absorption units are obtained. All of these classes can result in a variety of operating modes, depending on the direction of working fluid streams and on the internal heat exchange employed.

If cascade configurations are classified as seen in Figure 7.1, the basic cycles A and B must be distinguished because it is possible that, in each basic cycle, different working and absorber fluids are used. Therefore, each of the 16 different ways of coupling results in new classes of configurations.

7.2   OPERATING MODES OF TWO-STAGE ABSORPTION UNITS WITH SIX EXCHANGE UNITS AND TWO WORKING FLUID STREAMS (CLASSES OF FIGURE 7.1S AND T)

Figure 7.3 shows details of the coupling of six exchange units, ABCDEF, for the configuration of Figure 7.1t. Three complete basic cycles can be identified, ABED, BCFE, and ACFD. The first two cycles are chosen to be independent cycles, and the circulating working fluid flow rates are designated by m and n. This configuration represents as many as four types of two-stage absorption units, depending on whether the basic cycles are operated as a heat pump or as a heat transformer:

Type 1: Heat Pump/Heat Pump

   $m > 0, n > 0$

Type 2: Heat Transformer/Heat Transformer

   $m < 0, n < 0$

Type 3: Heat Transformer/Heat Pump

   $m < 0, n > 0$

Type 4: Heat Pump/Heat Transformer

   $m > 0, n < 0$

The concentrations and pressures in Figure 7.3 are chosen such that twice two exchange units fall into one temperature range. The devices for internal heat exchange may be turned on and off depending on the modes of operation, which will be discussed next.

To simplify the classification of modes of operation, it is assumed that the amount of heat transferred in any exchange unit is proportional to the net amount of working fluid undergoing phase transition in that exchange unit. For a quantitative evaluation of the efficiency, factors such as the temperature dependence of the latent heat, the composition dependence of the heat of absorption, the incomplete heat exchange in solution heat exchangers, and the heat capacity

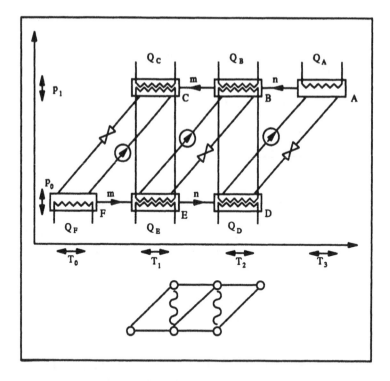

Figure 7.3   Absorption units with two independent cycles operating on the same pressure level. In operating range 2, the COP may be varied between 1.5 and 3.

of the liquid and gaseous working fluid have to be taken into account. However, detailed statements can be made about operating modes independent of these approximations. Only the precise location of the boundaries between operating ranges are affected by a more exact calculation.

The following values are obtained for the heating and cooling capacities Q_i of individual exchange units at temperature levels T_3 through T_0 (Q_i in units of heat of condensation):

$$Q_A = n \qquad\qquad Q_3 = n$$

$$Q_B = -n + m \qquad Q_2 = -2n + m$$

$$Q_C = -m \qquad\qquad Q_1 = n - 2m$$

$$Q_D = -n \qquad\qquad Q_0 = m$$

$$Q_E = n - m$$

$$Q_F = m$$

The fluid streams m and n are plotted on the ordinate and abscissa of a coordinate system, Figure 7.4. The boundaries for $Q_0 = 0$, $Q_1 = 0$, $Q_2 = 0$, and $Q_3 = 0$ are also displayed. Eight operating modes are found. Each is distinguished by a characteristic sign combination of Q_i. The combinations of signs of the respective operating ranges are summarized in Figure 7.5.

A comparison of Figure 7.5 with Figure 1.14, (Chapter One) shows that Figure 7.5 contains all reversible operating modes for four temperature levels. The basic absorption cycle can only be operated in modes 2 and 6 of Figure 7.5. By extending the basic cycle by one pair of

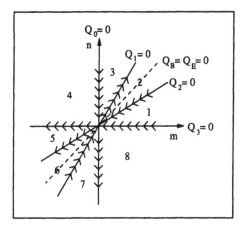

Figure 7.4 Operating ranges (see Figure 7.5) of absorption units according to Figures 7.3.

	①	②	③	④	⑤	⑥	⑦	⑧
Q_3	+	+	+	+	-	-	-	-
Q_2	+	-	-	-	-	+	+	+
Q_1	-	-	+	+	+	+	-	-
Q_0	+	+	+	-	-	-	-	+

Figure 7.5 Sign combinations for net amounts of heat exchanged in the absorption unit of Figure 7.3 for operating ranges 1 through 8 and temperatures T_3 through T_0 (see Figure 7.4).

exchange units, the configuration of Figure 7.3 can be operated in any of the modes of operation of Figure 7.5. When the pair C/F contains only pure working fluid, only one connection, made with either an expansion valve or with a pump, is required.

Operating Range 1
Heat is supplied to A and B, cooling is provided by F, and waste heat or useful heat is provided by C and E. The heat rejected by D can be used as additional heat supply to B. The heat ratio for cooling is given by:

$$\xi_c = \frac{\dot{Q}_0}{Q_2 + Q_3} = \frac{\dot{m}}{m - n} \tag{7.1}$$

It varies from $\xi_c = 1$ for $Q_3 = 0$ ($n = 0$) to $\xi_c = 2$ for $Q_2 = 0$ ($n = m/2$). As will be shown later, irreversibilities reduce the efficiencies by about 50% for the working pair NH_3/H_2O.
The heat ratio for heating is

$$\xi_H = \frac{|Q_1|}{Q_2 + Q_3} = \frac{2m - n}{m - n} \tag{7.2}$$

When B is at a higher temperature level than D, the heat rejected by D can be employed as useful heat.

Operating Range 2
Heat is supplied to A, cooling is provided by F, and useful heat is rejected at C, E, D, and B. The COP for cooling is

$$\xi_c = \frac{Q_0}{Q_3} = \frac{m}{n} \qquad\qquad (7.3)$$

It varies between $\xi_c = 2$ for $Q_2 = 0$ ($n = m/2$) and $\xi_c = 0.5$ for $Q_1 = 0$ ($n = 2m$)
The heat ratio for heating is

$$\xi_H = \frac{Q_1 + Q_2}{Q_3} = \frac{n + m}{n} \qquad\qquad (7.4)$$

The boundaries of operating range 2 and its center region represent the following special operating modes:

(a) Operating mode $Q_2 = 0$; i.e., $Q_B = -Q_D$; $n = m/2$

   The heat of absorption, rejected by D, can be used entirely for desorption in B. One-half of the working fluid condensed in C or evaporated in F is generated in A, while the other half is generated in B by internal heat exchange with D. The useful heat is obtained from C and E at the temperature level T_1. When the unit is used as a heat pump, a heat ratio of 3 is achieved. When the unit is used for refrigeration, the heat ratio is 2. This results in a 100% increase in efficiency when compared to a single-stage unit. The generator-absorber heat exchange displayed in Figure 4.28 (Chapter Four) may be considered as a special case of the configuration 7.3, for which the concentration differences of the two solution circuits are so large that the concentration of the strong absorbent in the high temperature solution circuit is identical with that of the weak absorbent in the lower temperature solution circuit. Depending on the flow rates in the two solution circuits, the two lines with the same concentration will be eliminated or replaced by a single line with a pump or a throttle.

(b) Operating mode $Q_B = 0$ and $Q_E = 0$; $n = m$

   There is no heat input or output in B and E. A basic cycle is obtained with a high absorption temperature. The heat ratio is 2 for heating and 1 for cooling. As is true for all other operating modes, exchange unit B may be operated as a rectifier.

(c) Operating mode $Q_1 = 0$; i.e., $Q_c = -Q_E$; $n = 2m$

   The condenser or absorber heat from C can be used entirely to generate working fluid vapor in B. The useful heat (or the waste heat from a refrigerant) is now rejected by D and D at the higher temperature T_2. The heat ratios are $\xi_H = 3/2$ or $\xi_c = 1/2$. The cooling capacity is now provided with about double the temperature lift when compared to the temperature lift of a single-stage unit.

In Figure 7.6, the latter three operating modes are summarized. It may be assumed that the evaporator capacity m is the same for all three operating modes and that the generator heat n is adjusted as required. In this case, increasing the temperature of the useful heat leads to an increased heating capacity and decreased heat ratio.

When examining the operating mode with a fixed generator heat n, then the heat ratio and the heating capacity increase with increasing evaporator capacity m. These relationships show how the configuration of Figure 7.3 is capable of meeting widely changing heating requirements such as varying heating capacities and outdoor temperatures. The configuration assures that the best possible heat ratio is obtained for any given condition. An example of a technical implementation of the configuration of Figure 7.3 is illustrated in Figure 7.7. It allows for a continuous transition between operating modes within and at the boundaries of range 2. The

Operating State	Working Fluid Streams	Driving Heat	Useful Heat	Temperature of useful heat
$\xi=3$	$n=m/2$	$Q_A=n=m/2$	$Q_C+Q_E=m+(m+n)=3/2m$	$1/3(2T_C+T_E)$
$\xi=2$	$n=m$	$Q_A=n=m$	$Q_C+Q_D=m+n=2m$	$1/2(T_C+T_D)$
$\xi=3/2$	$n=2m$	$Q_A=n=2m$	$Q_B+Q_D=(n-m)+n=3/2m$	$1/3(T_B+T_D)$

Figure 7.6 Some characteristic operating modes of an absorption unit according to Figure 7.3 in operating range 2.

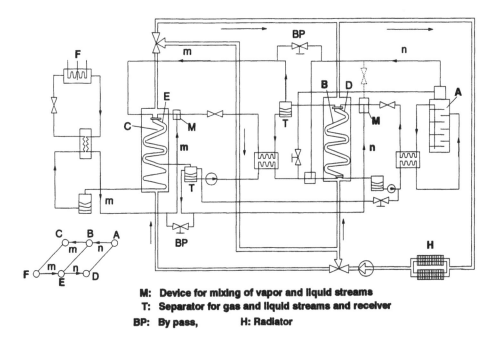

M: **Device for mixing of vapor and liquid streams**
T: **Separator for gas and liquid streams and receiver**
BP: **By pass,** H: **Radiator**

Figure 7.7 Detailed implementation of an absorption unit as seen in Figure 7.3.

internal heat and mass transfer occur in a tube-to-tube or tube-in-tube heat exchanger located inside a third tube which carries the water for external heat transfer. Depending on the amount of heat extracted from the triple heat exchangers BD and CE, a continuous transition between all three operating modes is accomplished. The heat ratio for heating for this unit ranges from 1.5 to 3. Due to irreversibilities, the heat ratio ranges from 1.25 to 2 for actual machinery for the working pair ammonia/water. The increase of the heat ratio of 2 of a single-stage absorption heat pump to the higher heat ratio of 3 which is achieved for increasing outdoor temperatures by the two-stage system shows that the two-stage system is clearly superior. In addition, the heat pump operation of the two-stage system permits an extension. Losses caused by cycling of a heat pump may be significantly reduced by a continuous adjustment of the heating capacity to varying operating conditions. In Figure 7.8, the heat ratio for heating is plotted as a function of n/m.

Operating Range 3

In operating range 3, heat is supplied to A, while cooling is provided by F and E. Useful or waste heat is rejected by B and D. The heat generated in C may be used internally for desorption in E or may be rejected, if possible. The heat ratios for cooling and heating are

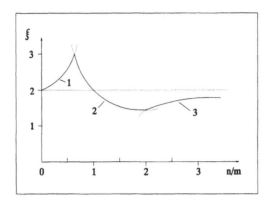

Figure 7.8   Heat ratio ξ_H of an absorption unit from Figure 7.3 in the operating ranges 1, 2, and 3.

$$(1)\, \xi_c = \frac{Q_1 + Q_0}{Q_3} = \frac{n-m}{n}, \qquad (2)\, \xi_H = \frac{|Q_2|}{Q_3} = \frac{2n-m}{n} \tag{7.5}$$

Figure 7.8 shows the heat ratio of a heat pump in Figure 7.3 for operating ranges 1, 2, and 3 as a function of the ratio n/m.

Operating Range 4

Operating range 4 describes a Type 3 heat pump transformer. The low temperature cycle BCFE is a heat transformer, and the high temperature cycle ABED is a heat pump. It may operate in several modes:

A.   Single-stage heat pump (m = 0),

B.   Heat transformer (n = 0),

C.   Combination of both (m ≠ 0, n ≠ 0)

The heat transformer is driven by heat input to C and E. The heat pump is driven by the heat supplied to A. Heat is rejected from both cycles by B and D. During periods of low heating loads and low outdoor temperatures, the unit can operate as a heat transformer. To increase the heating capacity or to raise the temperature of the useful heat, the heat pump operation of variable capacity may be added as needed.

In industrial applications this configuration may be used as a very efficient heat supply system for separation processes. Driving heat is supplied by A and low temperature heat is rejected by F. Useful heat is rejected by B and D at T_2 and used as input to a separation process. The output of the separation system may be available at T_1 and is supplied to C and E. With Q_1 equal to Q_2, the heat ratio $\xi = Q_2/Q_3$ amounts to 3.

Operating Ranges 5, 6, and 7

In the operating ranges 5, 6, and 7, both basic cycles operate as heat transformers. Operating mode 6 is especially important since the heat ratio can be varied continuously from 1/3 (n = m/2) to 1/2 (n = m) and to 2/3 (n = 2m), with maximum temperature lift for 1/3 and minimum temperature lift for 2/3.

Operating Range 8

The low temperature cycle operates in operating range 8 as a heat pump or refrigeration unit while the high temperature cycle operates as a heat transformer. This unit simultaneously produces cooling at F and upgraded heat at A at the temperature T_3. Heat is supplied to B and

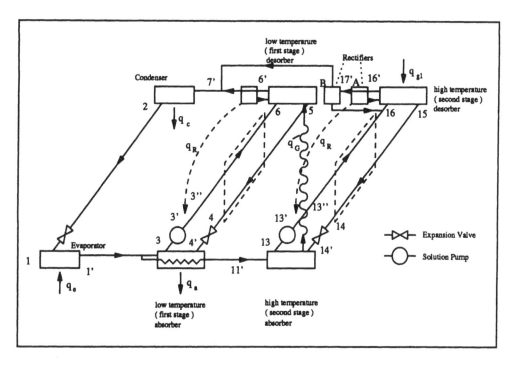

Figure 7.9 Schematic diagram of an absorption heat pump for Example 7.1.

D. The heat rejected at C and E can be either waste heat or useful heat, depending on the value of T_1. Both cycles operate independently of each other so that the ratio of heating capacity at T_3 to cooling capacity at T_0 can be adjusted for varying loads.

Example 7.1: Two-Stage Absorption Heat Pump Evaluation (Two Pressure Levels)

The two-stage absorption heat pump cycle shown in Figure 7.9 is to be evaluated. The unit operates in operating range 1, defined above. The heat of the second stage absorber (high temperature absorber) is completely used to drive the first stage (low temperature) desorber. The working fluid pair is ammonia and water. The effect of rectification is taken into consideration for both desorbers. The concentration of the vapor entering the condenser is 0.995. The evaporator temperature is 5°C, the lowest temperature in the absorber and the condenser temperature are 40°C, and the concentration of strong absorbent (solution low in ammonia concentration) is 0.36. The liquid leaving the condenser, both absorbers and both desorbers, is saturated. The effectiveness of the solution heat exchangers is 1.0. The temperatures in the high temperature absorber are 5 K higher than the temperatures in the low temperature desorber. The vapor entering the second stage absorber is preheated in the first stage absorber.

(a) What are the pressures, temperatures, and enthalpies of all relevant state points, including the vapor concentrations?

(b) What are the amounts of heat exchanged in the evaporator, the condenser, the absorbers, the desorbers, the rectifiers, and the solution heat exchangers? Base the calculation first on 1 kg of circulating refrigerant vapor and then balance the internal heat exchange.

(c) What is the COP of the system?

Solution: (a) The specifications for the low temperature side of the two-stage cycle, the first stage, are identical to those in Example 4.2. Therefore, all calculations for this system are not repeated, but the results will be used here. Nevertheless, all state points are indicated together, with the additional ones for the two-stage system on the process scheme of the two-stage ammonia/water absorption heat pump in Figure 7.9 and also on Figure 7.10, a schematic enthalpy-concentration diagram. All property data are summarized in Table 7.1. The points that are added for the second stage have the same number as the corresponding point of the first stage, but with ten added to it. For example, the absorber outlet is state point 3 for the first stage and 13 for the second stage.

The state points for the second stage are found to be analogous to the procedures described in previous examples for the corresponding points of a single-stage system.

The rectifier A of the second stage has to be explained in more detail. To use the heat of rectification for preheating of the solution leaving the high temperature absorber, the vapor 17′ cannot be colder than the temperature of point 13, 93°C. The vapor leaving the rectifier at 93°C must be in equilibrium with some liquid at the same pressure and temperature. From the h,x diagram, the concentration of this liquid is about 0.43. The vapor concentration in equilibrium with this liquid is 0.96, its enthalpy is 1480 kJ/kg, point 17′. Thus, it is assumed that the rectifier of the second stage produces vapor of the concentration 0.96. This vapor has to be rectified once more to the desired concentration of 0.995. This can be accomplished in several ways. The vapor can pass through the low temperature desorber and undergo some cooling and rectification while mixing with the first stage vapor. In this case, the first stage cycle has to be completely reevaluated since desorber heat, rectification heat, and, therefore, most other parameters would change. In addition, water would be transferred from the second stage to the first stage, requiring additional means to return the water to the second stage. In another way, the vapor would circumvent the first stage desorber and would be rectified in a third rectifier. The heat of rectification could be rejected to the condenser cooling water, and the reflux would return to the second stage desorber. This case is considered here for the sake of simplicity. The vapor leaving rectifier B has a concentration of 0.995, is essentially pure ammonia, and is represented by the properties of vapor at point 7′. (The assumptions regarding the rectifier are simplified and may lead to rectifier heats that are smaller than the actual value; a detailed description can be found in the literature.) The enthalpy at state point 11′, the vapor preheated in the first stage absorber, is found by using the ln(P)-h diagram for pure ammonia. The enthalpy change is determined between saturated vapor at 470 kPa and superheated vapor of the same pressure at 58°C, which is the saturated liquid temperature at the liquid inlet of the first stage absorber, point 4′. The enthalpy difference is 140 kJ/kg and is added to the vapor enthalpy of point 1′.

(b) The evaluation begins with the calculation of the first stage. All results are available from Example 4.2. The second stage is considered as a single-stage absorption heat pump using the same evaporator and

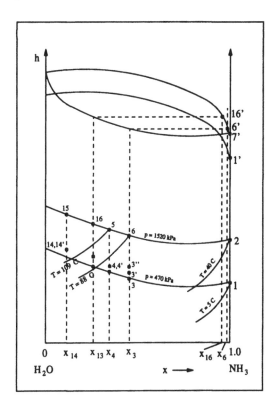

Figure 7.10 Schematic enthalpy-concentration diagram for Example 7.1.

**Table 7.1 Thermodynamic
Properties at the State Points**

	T(C)	P(kPa)	x	h(kJ/kg)
1	5	470	0.995	22
1'	5	470	0.995	1311
2	40	1520	0.995	194
3	40	470	0.46	-63
3''	47	1520	0.46	-34
4	60	1520	0.36	40
4'	58	470	0.36	31
5	109	1520	0.36	285
6	88	1520	0.46	173
6'	88	1520	0.970	1455
7'	40	1520	0.995	1340
11'	58	470	0.995	1451
13	93	470	0.19	268
13'	93	1520	0.19	269
13''	116	1520	0.19	376
14	114	1520	0.08	423
14'	137	1520	0.08	519
15	169	1520	0.08	675
16	146	1520	0.19	513
16'	146	1520	0.75	1854
17'	93	1520	0.96	1480

condenser as the first stage. The calculations are based on 1 kg of rectified vapor. The pump work w_{pss} is determined according to the usual procedure (the subscript ss stands for second stage). The flow rate f_{ss} through the pump is

$$f_{ss} = (x_{7'} - x_{15})/(x_{16} - x_{15})$$

$$f_{ss} = (0.995 - 0.08)/(0.19 - 0.08) = 8.3$$

The pump work amounts to:

$$w_{pss} = f_{ss}(P_2 - P_6)v = 8.3\,(1520\text{ kPa} - 470\text{ kPa})/800\text{ kgm}^{-3} = 10.9 \text{ kJ/kg}$$

With this, the enthalpy $h_{13'}$ of the fluid leaving the pump equals:

$$h_{13'} = h_{13} + w_{pss}/f_{ss} = (268\text{ kJ/kg}) + 10.9\text{ kJ/kg}/8.3 = 269 \text{ kJ/kg}$$

The amount of heat rejected by the rectifier A, Figure 7.9, is calculated first. The amount of reflux returning to the desorber as reflux v is

$$v_{ssA} = (x_{17'} - x_{16'})/(x_{16'} - x_{16})$$

$$v_{ssA} = (0.96 - 0.750)/(0.750 - .190) = 0.38$$

$$q_{rssA} = (h_{16'} - h_{17'}) + v_{ssA}(h_{16'} - h_{16})$$

$$q_{rssA} = (1854\text{ kJ/kg} - 1480\text{ kJ/kg}) + 0.38\,(1854\text{ kJ/kg} - 513\text{ kJ/kg})$$

$$q_{rssA} = 884 \text{ kJ/kg}$$

The heat of rectification is added to the solution leaving the pump and its enthalpy increases to $h_{13''} = h_{13'} + q_{rssA}/f_{ss} = 376$ kJ/kg. The temperature of the solution increased to 116°C, as seen in the h,x diagram in Figure 7.10. Rectifier B rejects the following amount of heat:

$$v_{ssB} = (x_{7'} - x_{16'})/(x_{16'} - x_{16})$$

$$v_{ssB} = (0.995 - 0.960)/(0.960 - .190) = 0.045$$

$$q_{rrsB} = (h_{17'} - h_{7'}) + v_{ssB}(h_{17'} - h_{16})$$

$$q_{rssB} = (1480\text{ kJ/kg} - 1340\text{ kJ/kg}) + 0.045\,(1480\text{ kJ/kg} - 513\text{ kJ/kg})$$

$$q_{rssB} = 184 \text{ kJ/kg}$$

The heat of rectification is added to the condenser cooling water. Now the solution heat exchanger can be evaluated:

$$q_{rsss} = f_{ss}(h_{16} - h_{13''}) = 8.3\,(513\text{ kJ/kg} - 376\text{ kJ/kg}) = 11137 \text{ kJ/kg}$$

$$q_{lsss} = (f_{ss} - 1)(h_{15} - h_{14}) = 7.3\,(675\text{ kJ/kg} - 423\text{ kJ/kg}) = 1840 \text{ kJ/kg}$$

q_{rsss} is the smaller amount of heat compared to q_{lsss}, and the enthalpy of point 14 has to be recalculated:

$$h_{14'} = h_{15} - q_{rsss}/(f_{ss} - 1) = 675 \text{ kJ/kg} - (1137 \text{ kJ/kg})/7.3 = 519 \text{ kJ/kg}$$

Now all relevant amounts of heat can be determined. From energy and mass balances for each of the heat exchangers follows:

$$q_{ass} = h_{11'} - h_{14'} + f_{ss}(h_{14'} - h_{13})$$

$$q_{ass} = 1451 \text{ kJ/kg} - 519 \text{ kJ/kg} + 8.3\,(519 \text{ kJ/kg} - 268 \text{ kJ/kg}) = 3015 \text{ kJ/kg}$$

$$q_{dss} = h_{7'} - h_{15} + f_{ss}(h_{15} - h_{16}) + q_{rssA}$$

$$q_{dss} = 1340 \text{ kJ/kg} - 675 \text{ kJ/kg} + 8.3\,(675 \text{ kJ/kg} - 513 \text{ kJ/kg}) + 884 \text{ kJ/kg} + 184 \text{ kJ/kg} = 3078 \text{ kJ/kg}$$

With this, all amounts of heat for the second stage are determined based on 1 kg of rectified ammonia vapor leaving the second stage. However, the high temperature absorber releases 3023 kJ/kg, and the low temperature desorber requires only 1950 kJ/kg. Thus, the amount of ammonia vapor produced in the second stage must be reduced by a factor of (1950 kJ/kg)/(3015 kJ/kg) = 0.647. All energy values and flow rates are scaled with this factor. The following amounts of heat and work are obtained for the entire cycle:

Pump work, w_p, w_{pss} 0.647 = 8.4 kJ/kg, 6.7 kJ/kg

Solution heat exchanger q_{rs}, q_{rsss} 0.647 = 1325 kJ/kg, 728 kJ/kg

Rectifier q_r, $(q_{rssA} + q_{rssB})$ 0.647 = 178 kJ/kg, 691 kJ/kg

Absorber heat q_a – 140 kJ/kg 0.647, q_{ass} 0.647 = 1839 kJ/kg, 1950 kJ/kg (The term 140 kJ/kg 0.647 represents the amount of heat required to preheat the vapor for the second stage absorber.)

Desorber heat q_d, q_{dss} 0.647 = 1950 kJ/kg, 1990 kJ/kg

Evaporator heat q_e 1.647 = 1839 kJ/kg

Condenser heat q_c 1.647 = 1887 kJ/kg

Checking the energy balance for the entire two-stage system yields:

$$q_d + q_e + w_p - q_c - q_a - q_{rssA}\; 0.647 = (1990 + 1839 + 15 - 1887 - 1839 - 119) \text{ kJ/kg} = -1 \text{ kJ/kg}$$

This result is not exactly equal to zero because of round-off errors. A deviation of less than 1% is acceptable.

(c) The COP amounts to

$$COP = q_e/(q_d) = 0.92$$

The pump work is neglected. The COP for the single-stage system was 0.57 (Example 4.2) and for the cycle with desorber/absorber heat exchange, 0.72 (Example 4.3). Thus, the two-stage cycle is clearly superior

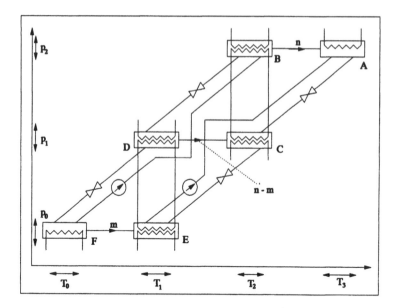

Figure 7.11   Absorption unit with two independent basic cycles on three different pressure levels.

to both. This results from a better match of the amounts of heat for the internal heat exchange. An inherent disadvantage compared to the single-stage and desorber/absorber heat exchange cycle is the requirement for two solution pumps. The performance of the two-stage system can be increased further by making better use of the heat of rectification, or by avoiding part of it by establishing water transfer from the low temperature absorber to the high temperature absorber through the vapor stream. The vapor to the high temperature absorber would not only be preheated, but would also pick up water and transport it to the high temperature desorber. In a practical system, controls have to be provided that ensure that the water content in both solution circuits is stable.

Figure 7.11 shows in more detail the configuration of Figure 7.1s. The exchange units B, D, and F can be condensers and evaporators. It is composed of two basic cycles and operates on three significant pressure levels. The same four types of absorption units are possible from this configuration with the same eight operating ranges corresponding to those described for the configuration of Figure 7.3. For given temperature levels T_0 through T_3, the two configurations shown in Figures 7.3 and 7.11 can be compared as follows: The device in Figure 7.11 requires higher pressures than that in Figure 7.3, which is a disadvantage when ammonia is the working fluid. The configuration in Figure 7.3 requires a wider solution field than the one shown in Figure 7.11. The configuration in Figure 7.3 can appropriately utilize the absorption pair NH_3/H_2O. The configuration in Figure 7.11 is more suitable for $H_2O/LiBr-H_2O$. The special operating mode, $Q_2 = 0$, of Figure 7.11 is used in commercial machinery and utilizes $H_2O/LiBr-H_2O$ as a working pair. This mode of operation is called a double-effect chiller (Figure 7.12). A process scheme of a technical version is shown in Figure 7.12, including the main components and the secondary components. The dashed line indicates the capacity control system.

The exchange units of the configurations from Figure 7.1s and t can be set to operate at five temperature levels instead of four. The internal heat exchange (wavy lines) is now between the exchange with C and D or B and E, respectively (Figures 7.13 and 7.14). The flows (in units of heat of condensation) exchanged by the configuration in Figure 7.13 are:

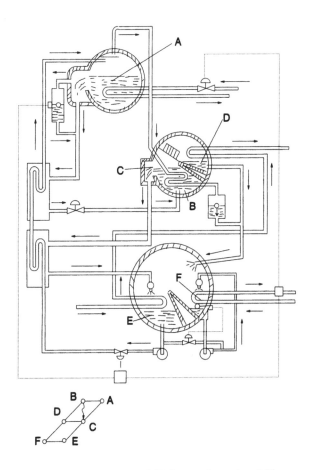

Figure 7.12 Two-stage LiBr/water absorption chiller.

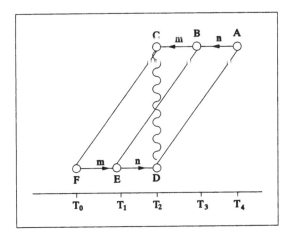

Figure 7.13 Two-stage absorption unit with two pressure levels.

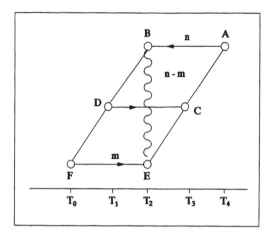

Figure 7.14 Two-stage absorption unit with three pressure levels.

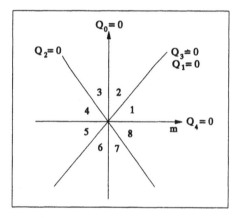

Figure 7.15 Operating ranges of units in Figure 7.13.

$$Q_A = n \qquad Q_4 = n$$
$$Q_B = n+m \qquad Q_3 = n+m$$
$$Q_C = -m \qquad Q_2 = -n-m$$
$$Q_D = -n \qquad Q_1 = n-m$$
$$Q_E = n-m \qquad Q_0 = m$$
$$Q_F = m$$

Figures 7.15 and 7.16 illustrate the operating ranges and the sign combinations of the heat flows Q_i for the temperature levels T_i. Identical operating modes and sign combinations are found for the configuration in Figure 7.14. The mode $Q_2 = 0$ ($m = -n$) is an example of a combination of a heat pump and a heat transformer as discussed previously as Type 3. Again, this configuration may be used as a heat supply system for a separation process with a large temperature difference between input and output.

	①	②	③	④	⑤	⑥	⑦	⑧
Q_4	+	+	+	+	-	-	-	-
Q_3	+	-	-	-	-	+	+	+
Q_2	-	-	-	+	+	+	+	-
Q_1	-	+	+	+	+	-	-	-
Q_0	+	+	-	-	-	-	+	+

Figure 7.16 Sign matrix of reversible solutions for five temperature levels.

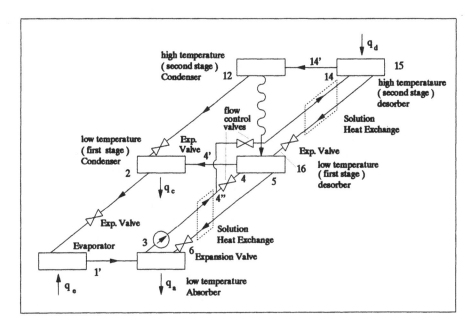

Figure 7.17 Process schematic diagram of a two-stage absorption heat pump with three pressure levels.

A comparison of Figure 7.16 with Figure 1.15 (Chapter One) shows that these combinations represent a subset of the reversible operating modes available with five temperature levels. There are at least three two-stage configurations required to obtain all reversible modes of operation for five temperature levels.

Example 7.2: Two-Stage Absorption Heat Pump Evaluation (Three Pressure Levels)

The two-stage absorption heat pump cycle shown in Figure 7.17 is to be evaluated. The unit operates in operating range 1. The heat released by the second stage condenser (high temperature condenser) is completely used to drive the first stage (low temperature) desorber. The working fluid pair is water and water/lithium bromide solution. The evaporator temperature is 5°C, the lowest temperature in the absorber and the condenser temperature are 40°C, and the concentration of strong absorbent (solution low in water content) is 0.63. The liquid leaving both condensers, the absorber, and both desorbers is saturated. The effectiveness of the solution heat exchangers is

1.0. The temperature in the high temperature condenser is 5 K higher than the highest temperature in the low temperature desorber. The condensate leaving the second stage condenser is expanded into the first stage condenser, and any flash gas is condensed. The solution pumped from the absorber is divided by means of flow control valves into two streams. The first stream enters the first stage desorber, and the second stream enters the second stage solution heat exchanger. Both desorbers produce a solution of the same concentration.

(a) What are the pressures, temperatures, and enthalpies of all relevant state points?

(b) What are the amounts of heat exchanged in the evaporator, the condensers, the absorber, the desorbers, and the solution heat exchangers?

(c) What is the COP of the system?

Solution: (a) The specifications for the low temperature side of the two-stage cycle, the first stage, are identical to those in Example 4.1. The calculations for this system are not repeated, but the results will be used here. All state points are indicated together, with the additional ones for the two-stage system on the process scheme in Figure 7.17 and also in Figure 7.18, a schematic enthalpy-concentration diagram. All property data are summarized in Table 7.2. The points that are added for the second stage have the same number as the corresponding points of the first stage, but with ten added to it. For example, the desorber outlet is state point 5 for the first stage and state point 15 for the second stage. The state points for the second stage are found to be analogous to the procedures described in previous examples for the corresponding points of a single-stage system.

(b) The evaluation begins with the calculation of the first stage. All results are available from Example 4.1, except for the pump work. The pump has to deliver fluid at a higher pressure, 101 kPa.

$$w_{p2} = f\left(P_2 - P_1\right) v_s$$

$$w_{p2} = 12.1 \left(101 \text{ kPa} - 0.87 \text{ kPa}\right) 1/1200 \text{ m}^3/\text{kg} = 1 \text{ kJ/kg}$$

The pump work is still negligible.

The second stage is considered a single stage absorption heat pump using the same evaporator and absorber as the first stage. The calculations are based on 1 kg of refrigerant vapor. The flow rate of solution into the high temperature desorber is determined according to the usual procedure. The subscript ss indicates variables for the second stage. The flow rate f_{ss} is the same as for the first stage, since the concentrations are the same. Next, the second stage solution heat exchanger is evaluated:

$$q_{1sss} = f\left(h_{14} - h_{4r}\right) = 12.1 \left(316 \text{ kJ/kg} - 172 \text{ kJ/kg}\right) = 1742 \text{ kJ/kg}$$

The heat capacity of the stream from 15 to 16, q_{rsss} amounts to:

$$q_{rsss} = (f - 1)\left(h_{15} - h_{16}\right) = 11.1 \left(355 \text{ kJ/kg} - 225 \text{ kJ/kg}\right) = 1443 \text{ kJ/kg}$$

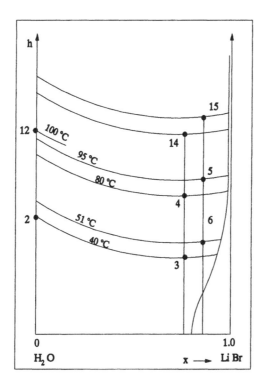

Figure 7.18 Schematic enthalpy-concentration diagram for Example 7.2.

Table 7.2 Thermodynamic Properties at the State Points

	T(C)	P(kPa)	x	h(kJ/kg)
1	5	0.7	0.0	21
1'	5	0.7	0.0	2510
2	40	7.4	0.0	168
3	40	0.7	0.58	106
4	79	7.4	0.58	189
4'	79	7.4	0.0	2650
1"	69	7.4	0.58	162
5	92	7.4	0.63	226
6	51	0.7	0.63	154
12	97	90	0.0	380
14	146	90	0.58	316
14'	146	90	0.0	2778
14"	132	90	0.58	291
15	160	90	0.63	355
16	92	90	0.63	225

It is assumed that the temperature and enthalpy of point 16 is identical to point 5. The enthalpy $h_{14''}$ has to be calculated:

$$h_{14''} = h_{4''} + q_{rsss}/f_{ss} = 172 \text{ kJ/kg} + (1443 \text{ kJ/kg})/12.1 = 291 \text{ kJ/kg}$$

The temperature at this point can be read from Figure 2.1 (Chapter Two) to 137°C. With this information the desorber heat can be evaluated:

$$q_{dss} = h_{14'} - h_{15} + f_{ss}\left(h_{15} - h_{14''}\right)$$

$$q_{dss} = 2778\ kJ/kg - 355\ kJ/kg + 12.1\left(355\ kJ/kg - 291\ kJ/kg\right) = 3197\ kJ/kg$$

The heat released in the high temperature condenser amounts to:

$$q_{css} = \left(h_{14'} - h_{12}\right) = \left(2778\ kJ/kg - 380\ kJ/kg\right) = 2398\ kJ/kg$$

The heat requirement for the first stage desorber is 3077 kJ/kg. Therefore, all flow rates and energies in the second stage have to multiplied by (3077 kJ/kg)/(2398 kJ/kg) = 1.283 to match the heat requirement of the first stage desorber.

Pump work, $w_p + w_{p2}$ 1.283 = 1.35 kJ/kg

Evaporator heat, q_e (1 + 1.283) = 5347 kJ/kg

Absorber heat, q_a (1 + 1.283) = 6705 kJ/kg

Condenser heat, q_c + 1.283 (380 − 168) = 2754 kJ/kg

The condenser heat is increased by the amount of heat released by the condensate from the second stage condenser.

Desorber heat, q_{dss} 1.283 = 4102 kJ/kg

The energy balance yields:

$$q_{dss} + q_e - q_c - q_a = 4102\ kJ/kg + 5347\ kJ/kg - 2754\ kJ/kg - 6705\ kJ/kg = -10\ kJ/kg$$

The deviation of -10 kJ/kg is acceptable.

(c) The COP amounts to:

$$COP = q_e/q_d = 1.30$$

The performance of the two-stage system is considerably higher than that of a single-stage system with a COP of 0.76.

7.3 INTERNAL HEAT EXCHANGE

The configurations in Figure 7.1a to r may be discussed in a similar way as previously discussed for Figure 7.1t and s. By shifting pressures and concentrations, internal heat exchange between two or more exchange units becomes possible. The Appendix shows tables of the options for internal heat exchange for the configurations a to r of Figure 7.1. In Figures 7.19 and 7.20, a graphical representation of the options for internal heat exchange for the configurations 7.1c and 7.1d is presented as an example. The term B_iA_j represents internal heat exchange that takes place between the exchange unit i of cycle B and the exchange unit j of cycle A. A plus sign and a minus sign occurring in the parentheses means that the flow in one cycle has a positive direction and in the other cycle a negative direction. The combination (+ −) also represents the case (− +) for which the direction of the heat flow for the internal heat exchange is reversed. If the signs are (+ +), both operate in the same direction. The combination (+ +) also represents the case (− −) with reversed heat flow (wavy line).

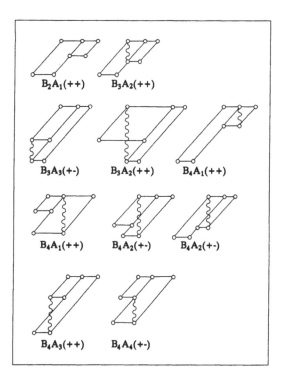

$B_2A_1(++)$ $B_3A_2(++)$

$B_3A_3(+-)$ $B_3A_2(++)$ $B_4A_1(++)$

$B_4A_1(++)$ $B_4A_2(+-)$ $B_4A_2(+-)$

$B_4A_3(++)$ $B_4A_4(+-)$

Figure 7.19 Internal heat exchange on one temperature level for configurations of Figure 7.1c. Identical signs indicate that the working fluid circulates in the same direction for both basic cycles; different signs indicate opposite flow directions.

All configurations in Figure 7.1a to r can operate in all eight reversible operating ranges, as in Figures 1.14 (Chapter One) and 7.5. For five temperature levels, only a subset of the reversible modes of Figure 1.15 can be realized by these two-stage machines. With internal heat exchange, either the efficiency or the temperature lift can be increased. A large temperature lift can be useful for obtaining very low refrigeration temperatures or combined production of refrigeration and useful heat.

In Figure 7.19, the internal heat exchange is performed on one temperature level. In Figure 7.20, it occurs simultaneously on two temperature levels.

The internal heat exchange with only one temperature level can always be complete by choosing the fluid flow rates in the two cycles appropriately. Complete internal heat exchange on two temperature levels is generally not possible. Only one parameter can be chosen independently when only two fluid streams are involved. Complete internal heat exchange on two temperature levels can always be obtained when an additional exchange unit is incorporated within the original configurations of Figures 7.1a to r in such a way that a third independent working fluid stream is created, Figure 7.2. The additional exchange unit is represented by a full circle. The new configurations of Figure 7.21 represent a subset of three-stage units with three independent working fluid streams. This offers a new degree of freedom. Since the configurations of Figure 7.1a to r fulfill, in many instances, the requirements for complete internal heat exchange on two temperature levels, only a relatively small exchange of heat and working fluid is necessary in the additional exchange unit. The added exchange unit is therefore expected to be small compared to the other units and can be designed accordingly. The exchange of heat or working fluid within the additional exchange unit, changes of its temperature or pressure level, and combinations of these variables can be used for control purposes.

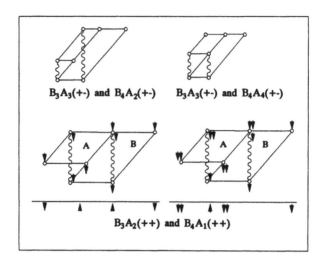

Figure 7.20 Internal heat exchange on two temperature levels for configurations of Figure 7.1c and d.

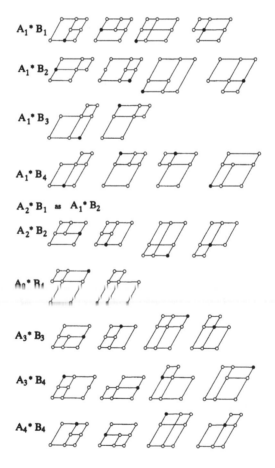

Figure 7.21 Extensions of configurations of Figure 7.1 by the addition of one, possibly small, exchange unit with three independent working fluid streams.

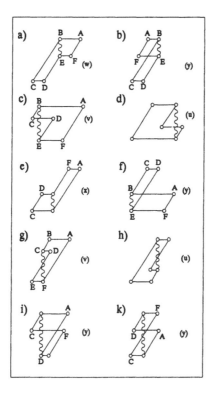

Figure 7.22 Internal heat exchange for the units of Figure 7.1u to y.

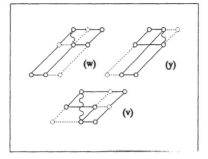

Figure 7.23 Complete internal heat exchange accomplished by a small additional exchange unit (dashed lines).

The configurations of Figure 7.1u to y operating with only one working fluid stream also offer a number of opportunities for internal heat exchange (Figure 7.22). This heat exchange is usually not complete, but completion is again possible with the addition of a small exchange unit. This completion can be carried out in three ways, as illustrated in Figure 7.23 for the configurations of Figure 7.1v, w, and y. By the addition of a small auxiliary component, the transfer of excess heat at the temperature level of the internal heat exchange is shifted to the auxiliary component. In the configurations of Figure 7.22a, b, e, and f, the exchange units coupled with internal heat exchange can be shifted parallel to the vapor pressure lines to higher or lower pressure levels without affecting the temperature levels of the heat exchange with outside heat reservoirs. Therefore, despite given temperature levels, the pressures can be chosen almost independently of temperature levels. This was not possible with single-stage units.

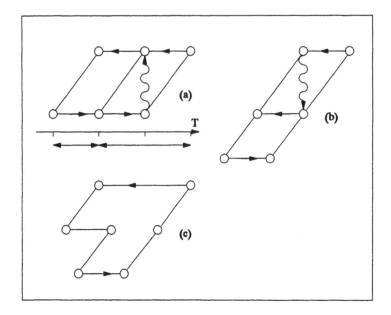

Figure 7.24   Absorption units with doubled cooling capacity (six exchange units).

7.4   HIGH EFFICIENCIES (HEAT RATIOS)

There are three configurations with six exchange units with a cooling efficiency ξ_k that is theoretically twice as high as that for the single-stage absorption unit. The configuration in Figure 7.24b, using the working fluids $H_2O/LiBr$-H_2O, is known as a double-effect unit. The double-effect unit is today widely used in Japan (see also Figure 7.12). When the fluid pairs NH_3/H_2O were used in the double-effect unit, the high side pressure would reach 80 bar. The configuration in Figure 7.24a is also a double-effect cycle with NH_3/H_2O; it requires a maximum pressure of about 25 bar, similar to that of a single-stage unit. The third cycle has both an evaporator and a desorber to produce refrigeration capacity. Therefore, one unit of working fluid produces about two units of refrigeration. All three configurations use either the condenser heat or the absorber heat of a first cycle as driving heat for a second cycle. The configuration of Figure 7.24c was created by applying Rule 3a to the configuration of Figure 7.1d, for example. This heat exchange between the first and second cycle occurs in the eliminated exchange unit. Another way of looking into the double-effect phenomena of the cycle in 7.24c is as follows. The working fluid produced in the generator is used twice for refrigeration.

Many of the configurations of Figure 7.1a to r which utilize seven exchange units can achieve the same doubled efficiency of $\xi_K = 2$.

The following three configurations are capable of tripling ξ_K. For all three configurations in Figure 7.25, the condenser heat and the absorber heat of a first cycle are used to drive a second cycle. The first cycle simultaneously provides cooling in an evaporator or desorber, respectively. As was shown in Chapter Six, the same principle can be employed for configurations that are coupled by heat exchange only (cascades). These configurations contain eight exchange units. As working fluids, combinations of Na, K, Li -hydroxides can be used simultaneously for both cycles, or $LiBr/H_2O$ for the small cycle and hydroxides for the large one. It should be noted that the double-effect units can be extended to triple-effect and even quadruple-effect machines by adding two exchange units. These units can be designed as cascades, too.

All configurations in Figure 7.25a have in common the fact that the heat of absorption or the heat of condensation of the added basic cycles is used as driving energy for the double-

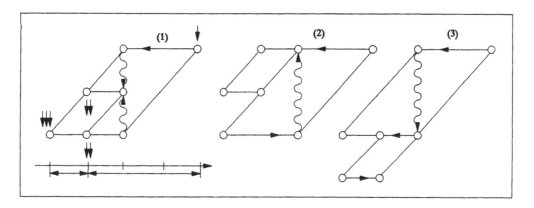

Figure 7.25   Absorption units with a tripled cooling capacity (seven exchange units).

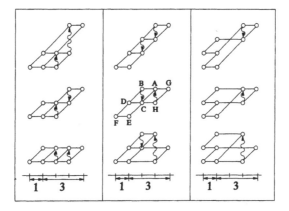

Figure 7.25a   Triple-effect configurations with eight exchange units.

effect unit. When the temperatures of the cooling capacity and of the waste heat are fixed, then the temperature of the generator heat is increased to about 200°C. This is necessary to improve the efficiency of a double-effect unit to that of a triple-effect unit. In case low temperature cooling water is available, the generator temperature can be lowered.

Especially noteworthy is the second configuration of the second column in Figure 7.25a. The conventional double-effect LiBr/H₂O unit ABCDEF is extended by one solution circuit GH yielding an additional basic cycle BCHG. The additional solution circuit may be operated with aqueous hydroxides. The high side pressure remains below 1 bar, and the hydroxide is used only in the range between 150 to 200°C. The high temperature generator and solution heat exchanger are relatively small and may be manufactured from Ni-alloys or carbon.

Figure 7.25b shows four quadruple-effect configurations with eight exchange units. Each unit of heat supplied to the high temperature generator is used four times to generate refrigerant for the evaporator. The generator temperature has to be increased again to about 260°C. Thus, it seems to be a worthwhile effort to develop working fluids and heat exchanger materials for this temperature range.

Two three-stage configurations employing eight exchange units are shown in Figure 7.26, which utilize the generator heat four times. Figure 7.27 summarizes the minimum number of necessary exchange units E to obtain a certain number of effects.

Figures 7.28 to 7.30 show configurations of double-, triple-, and quadruple-effect machines superimposed on the vapor pressure diagrams for NH₃/H₂O-LiBr and NH₃/H₂O. The units are driven with heat of 180°C and provide cooling at 10°C. The increase of the efficiency by

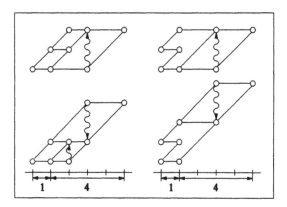

Figure 7.25b   Quadruple-effect configurations with eight exchange units.

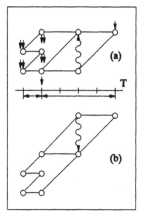

Figure 7.26   Absorption units with quadrupled cooling capacity.

Number of effects	Number of Exchange Units	Number of Stages	Figure
Single-effect	E=4	s=1	7.3
Double-effect	E=6	s=2	7.24
Triple-effect	E=7	s=2	7.25
Quadruple-effect	E=8	s=3	7.26

Figure 7.27   The minimum number of exchange units required for a given number of stages and for a given heat supply. (E, number of exchange units, s, number of stages).

adding effects is accompanied by a drop in the waste heat temperature from 50°C to 30°C. The unit shown in Figure 7.28 operates as a double-effect machine which may be air cooled. The unit in Figure 7.30 requires a wet cooling tower or cooling water. For the configuration of Figure 7.29, the waste heat temperature can be raised slightly by either installing a third liquid pump between exchange units C/D or by using counter-current heat exchange between C and B. Figure 7.31 indicates that the "quadruple effect" configurations of Figure 7.30, as well as

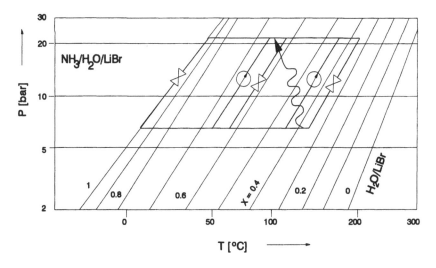

Figure 7.28 Absorption cycle with doubled cooling capacity.

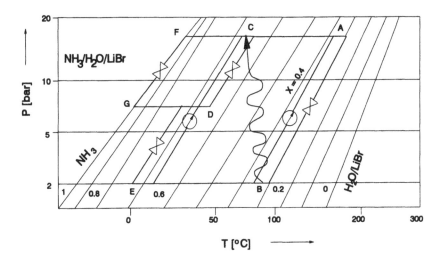

Figure 7.29 Absorption cycle with tripled cooling capacity.

all previous configurations, can be operated with NH$_3$/H$_2$O, but high water-vapor concentrations of the ammonia vapor leaving the generator are entailed.

7.5 LARGE TEMPERATURE LIFTS

Frequently, it is necessary to lower the refrigeration temperature for given temperature levels of driving heat and of waste heat. Heat-pumping applications present a similar task. For given temperature levels of the driving heat and the low temperature heat, the level of the useful heat should be increased. For the three configurations of Figure 7.32, the cooling capacity is obtained at a temperature level two temperature differences (one unit in temperature difference: [$T_{\text{driving heat}} - T_{\text{waste heat}}$]) below the level of the waste heat. The configuration of Figure 7.32c, using ammonia/water as the working pair, has been used for industrial applications. This unit has two generators; one cycle produces the cooling capacity, and the second cycle lifts the waste heat from the condenser or the absorber of the first cycle to the temperature of the surroundings for heat rejection. The heat exchange may be envisioned as being accomplished

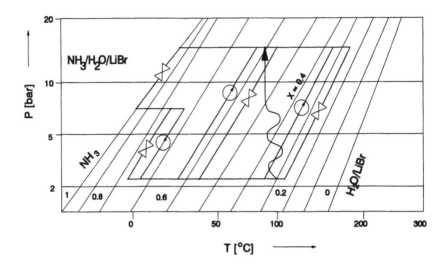

Figure 7.30 Absorption process with quadrupled cooling capacity.

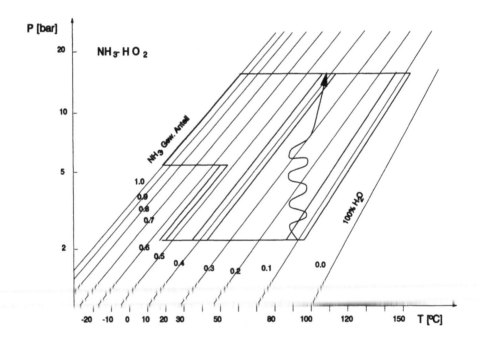

Figure 7.31 Absorption process with quadrupled cooling capacity.

in the eliminated exchange unit. The configuration of Figure 7.32a produces the same result, using only one generator and without requiring a second high pressure level.

Operating modes in which a lift across two units of temperature difference are accomplished can also be obtained with configurations of Figure 7.1a to r with seven exchange units. Out of these, there are three configurations which provide cooling at a temperature level three units below that of the waste heat Figure 7.33. All three configurations have a first cycle that lifts the condenser and absorber heat of a second cycle to the temperature level of the surroundings. The second cycle delivers the cooling capacity.

The three-stage configurations shown in Figure 7.34 with eight exchange units achieve refrigeration four temperature differences below the waste heat level. Figure 7.35 shows

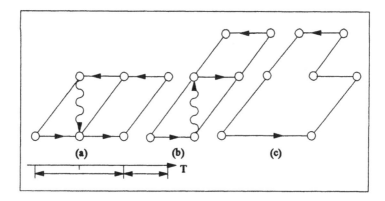

Figure 7.32 Examples of cycles that lift heat over two units of the temperature difference (six exchange units).

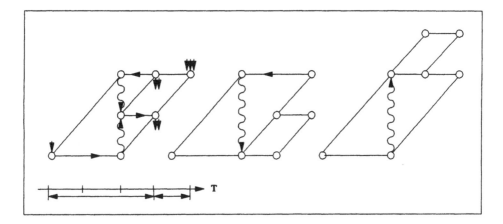

Figure 7.33 Heat pumping across three temperature differences (seven exchange units).

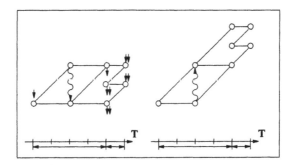

Figure 7.34 Heat pumping across four temperature differences (eight exchange units).

measured efficiencies of a LiBr/H_2O absorption heat pump according to Figure 7.32b with doubled temperature lift.

7.6 HEAT TRANSFORMER

When the directions of all heat and fluid streams are reversed in the configurations of Figures 7.24 to 7.26 and 7.32 to 7.34, heat transformers are obtained. The units described in Section 7.4, above, raise the temperature level of the useful heat. The high temperature lift of these units produces efficiencies that are lower than the efficiencies of single-stage heat transformers.

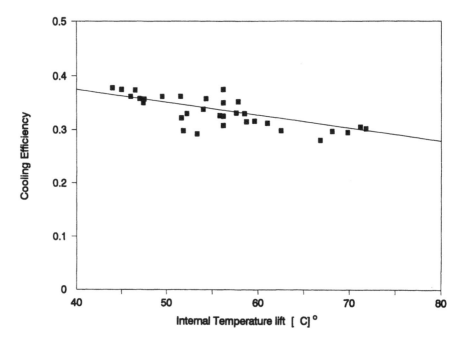

Figure 7.35   Measured cooling efficiency of the double-lift unit according to Figure 7.32b.

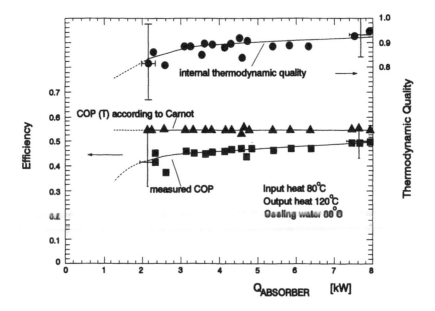

Figure 7.36   Measured efficiency and thermodynamic quality of a single-stage heat transformer for full and part load.

The units described in Section 7.5, above, yield high efficiencies, but only small temperature lifts.

Figure 7.36 shows the measured efficiency of a single-stage heat transformer operating with LiBr/H_2O as a function of the heating capacity (lower curve). The measured efficiency of 0.48 is rather close to the theoretical value of 0.5. An excellent part-load behavior can be noticed. The second line shows the efficiency of a completely reversible machine. The ratio of the two values (upper line) can be termed Second Law efficiency or thermodynamic quality

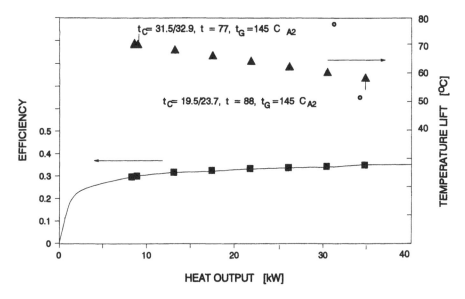

Figure 7.37 Measured efficiency and temperature lift for a double-lift heat transformer for full and part load.

factor. The figure shows that the efficiency of this experimental heat transformer reaches about 90% of the limit set by the Second Law. Since the Carnot efficiency is based on internal temperatures, the quality factor does include the irreversibilities within the machine, but not those for the heat exchange with external heat transfer fluids. If these were included, the quality factor decreases for increasing heating capacity and varies between 0.85 and 0.75. In Figure 7.37, the measured efficiency of a double-lift heat transformer operating with LiBr/H_2O according to the configuration in Figure 7.24b (with reversed heat and flow directions) is plotted as a function of the useful heat output. The right-hand scale shows the temperature lift which varies between 60°C and 70°C. Due to the large temperature lift, the efficiency is reduced to about 0.32. Again, a good part-load behavior can be noticed.

7.7 HEAT PUMP TRANSFORMER

In Section 7.2, heat pump transformers were described for the configurations of Figure 7.1s and t. Any of the configurations in Figure 7.1a to r can be operated as heat pump transformers with opposite working fluid directions in the two basic cycles. Figure 7.38 shows a heat pump transformer design in which the heat pump operates according to the configuration in Figure 7.1v and the heat transformer according to the configuration of Figure 7.1u. The two-stage cycles have four exchange units in common. This unit can be used as a high efficiency heat supply system for separation processes. Driving heat is supplied at T_3, and heat is rejected at T_2 and supplied to the separation process. The waste heat of the separation process is transferred back into the heat supply system at T_1. Finally, waste heat is rejected at T_0. If the separation process only degrades heat but has no net heat consumption ($Q_2 = Q_1$), then the theoretical ratios of the four amounts of heat for the device in Figure 7.38 are $Q_3: Q_2: Q_1: Q_0$ = 1:5:5:1. The measured value of Q_3, Q_2 is slightly above 4.0, Figure 7.39.

7.8 HEATING AND COOLING CAPABILITIES

When considering practical applications for heating and cooling systems, the economic focus is usually on either the cooling or the heating capacity. Some configurations combine both heating and cooling capabilities in one unit. This is possible when the waste heat of the refrigeration system is on a temperature level that is sufficiently high for heating.

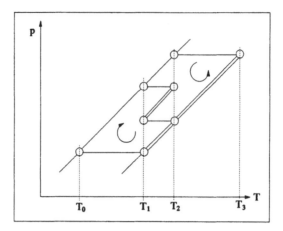

Figure 7.38 Heat pump transformer composed of a two-stage heat pump (double-effect) and a two-stage heat transformer (double-effect).

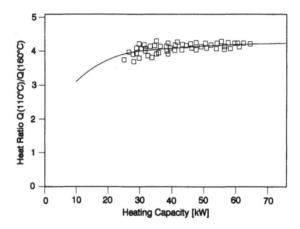

Figure 7.39 Measured heat ratio q_3/q_4 for the heat pump transformer of Figure 7.38.

Such a dual purpose system is shown in Figure 7.40 and may be called a double-lift, double-effect machine. So far, double-effect units (Figure 7.12) are used for cooling in summer During the heating season, the heat pump effect is abandoned. The steam involved in generator A (Figure 7.40), condenses in the auxiliary condenser 30 and returns directly into the generator when valve 29 is closed.

Figure 7.40 shows a modification of the double-effect unit. With only a small investment, the heat pump can be used for heating, especially during spring and fall. In summer, in addition to the cooling capacity, useful heat at a high temperature level of 60°C to 90°C can be obtained. The following four modes of operation are possible with suitable valve settings:

1. Double-Effect Operation
Heat is supplied to generator A. Valve 29 is open. Steam condenses in B. This generates additional steam in C, which condenses in D, and supplies heat to 7. The condensate flows through pipe 9 to the evaporator F. Pump 10 is used to recirculate the working fluid, water. Chilled water is produced in 12. The evaporated water is absorbed in E rejecting the absorber heat to 13. Pump 19 recirculates the weak absorbent to A and C. Components 16 and 18 are solution heat exchangers.

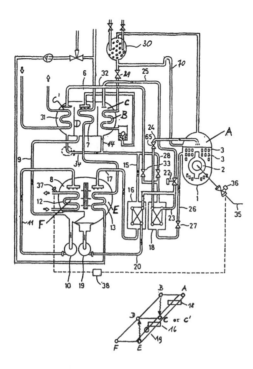

Figure 7.40 Two-stage LiBr/water absorption unit for cooling and heating.

2. Cooling and Useful Heat at High Temperatures

When it is desirable to provide useful heat and cooling capacity, part of the vapor stream from A is condensed in 30. The condensate, however, does not return to A, but proceeds through B and D to F where it provides refrigeration. If all vapor is condensed in 30, the unit operates as a single-effect machine. Cooling is provided between 5°C and 10°C for air conditioning and, at the same time, useful heat is generated from about 90°C to 100°C. This operating mode produces heat which might be used in a facility such as a hotel during times of reduced cooling needs.

3. Cooling and Useful Heat on a Moderate Temperature Level

When cold weather conditions require increased heating capacity, the absorber heat from E of about 30°C to 35°C can be transformed to a higher temperature of 50°C to 60°C. Components E and D are brought into internal heat exchange by disconnecting the cooling tower and by producing a "short circuit" between the heat exchange coils 7 and 13. Component D, used as a condenser in the first operating mode, is now used as an evaporator. At the same time, exchange unit C′ works as an absorber. In component 31, additional useful heat is generated. This operating mode enables heating in winter and cooling in summer when high waste heat temperatures are required (air-cooled systems).

4. Direct Heating

When outdoor temperatures drop, the heat pump operation can be suspended by closing valve 29. This switches the unit to direct heating. Since the addition of LiBr to the evaporator prevents water from freezing, this change-over point may be below 0°C.

In summary, Figure 7.40 shows a LiBr-H₂O air-conditioning unit capable of both heat pumping and cooling operations, by varying the heat exchange between the exchange units B and C or E and D, respectively.

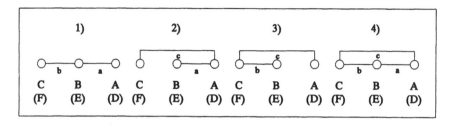

Figure 7.41   Configurations for connecting tubing containing refrigerant vapor.

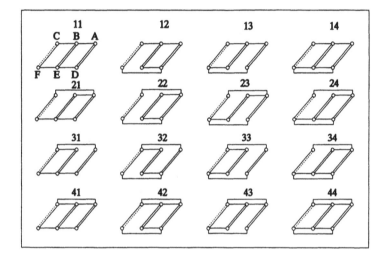

Figure 7.42   Variations of the configuration of Figure 7.1t.

7.9 CONNECTIONS FOR WORKING FLUID AND SOLUTION STREAMS

All vapor and solution circuits in Figure 7.1 are shown as plain straight lines. However, within each class several options exist for arranging the connections between exchange units, pumps, and expansion valves. When three exchange units operate on the same pressure level, there are four possible connections, as indicated in Figure 7.41. Vapor lines for the device in Figure 7.3 can be arranged in 16 variations, as shown functioning in Figure 7.42. Bypasses may avoid unnecessary pressure drop. When working fluid vapor is conducted through an exchange unit, gains in heating or cooling capacity or rectification may result

A configuration with three pressure levels many contain a multitude of connections for liquid working fluid (examples in Figure 7.43) and liquid working fluid/absorbent mixtures (examples in Figure 7.44). The double-effect machine is commercialized with solution circuitry called serial or parallel flow, depending on how the solution is circulating between the first and second generator.

Solution circuits can be merged when they are located side-by-side and when the concentration of the strong absorbent in one solution circuit is equal to the concentration of the weak absorbent in the other solution circuit. This is pictured in Figure 7.45 for pairs of exchange units A/C, B/D, and E/F.

On the left of Figure 7.45, two or three separate solution circuits are presented. The left line of a solution circuit contains relatively weak absorbent, and the right line contains relatively strong absorbent. On the right side of Figure 7.45, the solution circuits obtained by merging the circuits on the left are shown. Several variations are possible, for which only three examples are shown for a and b. If the directions of the fluid streams in the lines to be merged are inverted, the pump must be replaced by an expansion valve or vice versa.

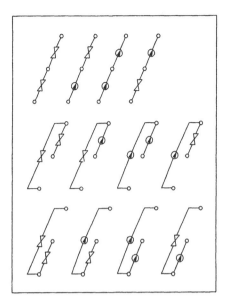

Figure 7.43 Variations of configurations for connecting tubing containing liquid working fluid.

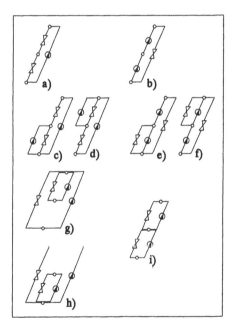

Figure 7.44 Variations of configurations for solution circuits between three exchange units.

Whenever the fluid streams oppose each other and are of the same magnitude, the line can be eliminated.

When two solution circuits are merged, the combined circuits can contain two exchange units operating at the same pressure level and connected by horizontal lines carrying solution, as seen in Figure 7.45a. These exchange units may remain separate. However, when a compact design is desirable, the units may be merged to one exchange unit, covering the entire temperature and composition gradient along the solution stream at constant pressure. Internal heat exchange, which is possible between individual exchange units, merges into heat exchange between the cold and warm end of combined exchange units at different pressure

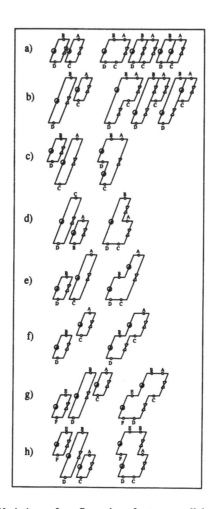

Figure 7.45   Variations of configurations for two parallel solution circuits.

levels. For this situation, the term "overlapping temperatures", as discussed in Chapter 4.5, is used. Therefore, the concepts of the "GAX cycles" (generator-absorber heat exchange) and "branched GAX cycle" are variations of the configuration in Figure 7.1t with variations of the solution circuits shown in Figure 7.45a.

7.10   SWITCHING BETWEEN CONFIGURATIONS OF DIFFERENT CLASSES

Discussion of the configurations of Figure 7.1 has shown how operating modes can be adjusted to varying conditions by changing the ratios and directions of fluid streams. Further adjustments can be made by providing additional connections between exchange units fitted with valves. Such arrangements allow "switching" between different classes of Figure 7.1a to z.

First, only connections carrying working fluid vapor are considered. Figure 7.46 illustrates which connections and valves are required to switch arbitrarily between the configurations of Figure 7.1u to z (even t can be included without modification of the solution circuits). The configuration of Figure 7.46 can be transformed into any of the seven classes in Figure 7.47.

The symmetrical representation in Figure 7.48 is identical to that of Figure 7.46, except that the solution circuits are shown as double lines. If switching is desired between fewer classes, a reduced number of vapor connections with valves is necessary. Even switching between certain classes of configurations with internal heat exchange is possible, as illustrated in Figure 7.49. This configuration can be transformed into the configuration of Figure 7.50 by

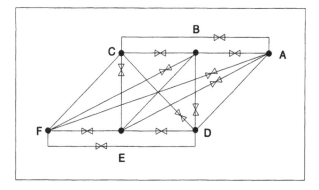

Figure 7.46 Switching between the seven classes of Figure 7.42 using valves in the vapor lines.

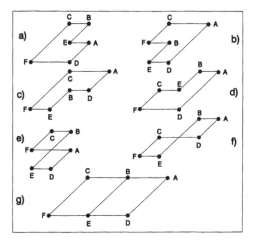

Figure 7.47 Examples of configurations resulting from switching of valves in Figure 7.46.

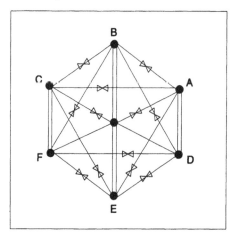

Figure 7.48 Switching between the seven classes of Figure 7.47 by valves in the vapor lines.

closing valve v_1 and by opening valves v_2 and v_3. By opening v_1 and closing v_2 and v_3, the configuration of Figure 7.51 is obtained. The configurations are distinguished by different temperature lifts resulting in different COPs.

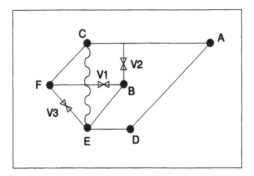

Figure 7.49   Switching between the two variations of Figures 7.50 and 7.51.

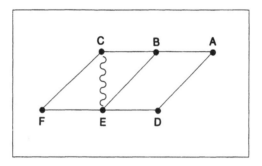

Figure 7.50   Two-stage absorption configuration.

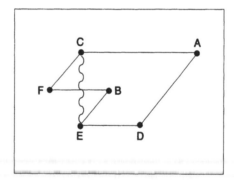

Figure 7.51   Two-stage absorption configuration with three pressure levels.

In addition, the 18 classes of Figure 7.1a to r can be divided into three sets, S_1 though S_3, of six classes. Each class in one set can be transformed into another class by switching valves in gas lines. Consequently, only the pressure levels on which certain exchange units work are changed. The three sets include the following classes: S_1(a,b,f,g,q,r), S_2(c,d,e,l,o,p), and S_3(h,i,j,k,m,n). The common characteristic of each of these sets is the relative location of the solution circuit connecting three exchange units with regard to those solution circuits that connect only two exchange units. Figure 7.52 illustrates the six classes of the group S_2(c,d,e,l,o,p) in which the solution circuit containing three exchange units is located in between the other two solution circuits. Figure 7.53 displays a combination of all configurations of Figure 7.52. For all operating modes to be realized, any four valves of the eight shown must be open. The four remaining valves are closed. This concept can also be applied to groups S_1 and S_3.

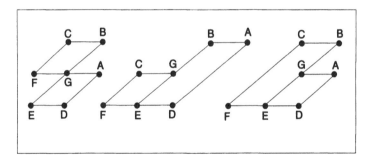

Figure 7.52 Six classes with the solution circuit containing three exchange units located in between the other two solution circuits.

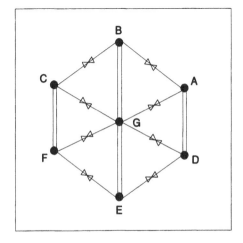

Figure 7.53 Switching between the six classes of Figure 7.52 by valves in the vapor lines.

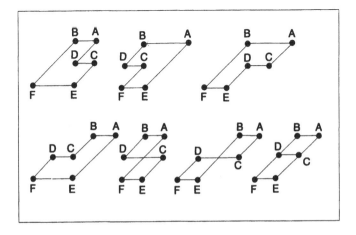

Figure 7.54 Six classes of configurations according to Figure 7.1u to z and the classes of Figure 7.1s.

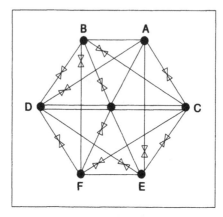

Figure 7.55 Switching between the seven classes of Figure 7.54 by valves in liquid lines.

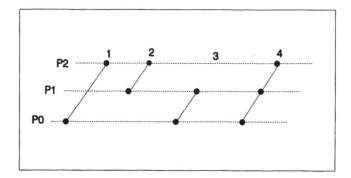

Figure 7.56 Pairs and triplets of exchange units connected by liquid lines.

Switching between classes has, so far, been obtained by opening and closing valves in pipes for working fluid vapor. Switching between classes can also be accomplished by opening or closing valves in pipes carrying liquid fluids.

Figure 7.54 shows the six classes of Figure 7.1u to z and the classes of 7.1s. Figure 7.55 shows how these classes can be transformed into each other by switching valves in liquid pipes. When valves in solution circuits are switched, usually two valves have to be opened or closed simultaneously. The switching of liquid lines may be necessary when low vapor pressures of the working fluid prevent the switching of vapor lines. These are often omitted to avoid pressure drops and to obtain a compact design (for example $H_2O/LiBr$ or $CH_3OH/LiBr$).

In the configurations of Figure 7.1a to r, again three groups are found which can be transformed into each other by switching valves. These groups are characterized by the pressure level on which three exchange units operate. These sets are identified as: S_4(a,b,c,d,h,i), S_5(e,f,g,k,j,l), and S_6(m,n,o,p,q,r). That switching is only possible between certain classes and can be attributed to a symmetry property, which manifests itself most clearly by the application of permutation operators. This procedure establishes sets in the mathematical sense (Section 7.11, below).

7.11 SET PROPERTIES

The configurations of Figure 7.1u to z as well as each of the six classes within the sets S_1-S_6, have set properties in the mathematical sense. The change of the relative locations within the

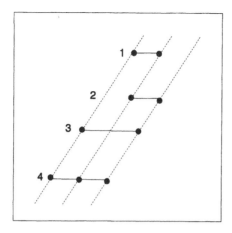

Figure 7.57 Pairs and triplets of exchange units connected by vapor lines.

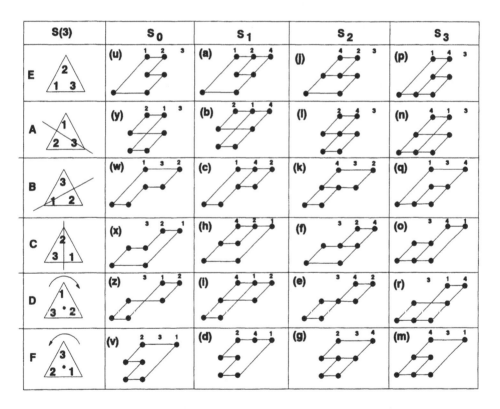

Figure 7.58 Application of the permutation operators A to F on the solution circuits in Figure 7.56.

P,T-diagram of objects which consist of combinations of exchange units are isomorph in relation to the six elements of the permutation set S(3). The set elements of S(3) permitted the arrangement of combinations of exchange units which are connected diagonally (liquid lines) or horizontally (vapor lines). Figures 7.56 and 7.57 represent four elements which act as building blocks of the configurations of Figure 7.1a to t. The sets S_0 to S_6 are composed of a combination of three elements (out of the four given) and of all permutations of the relative arrangements of these elements. The result of the application of all six permutation operators

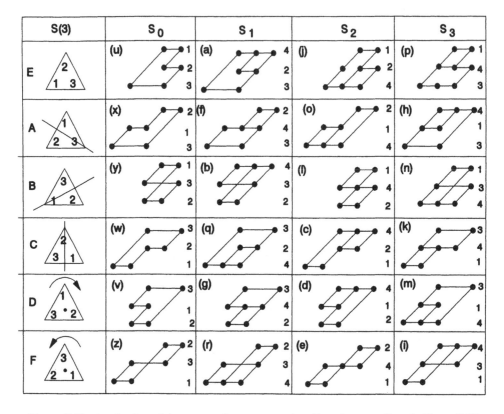

Figure 7.59 Application of the permutation operators A to F on the vapor lines in Figure 7.57.

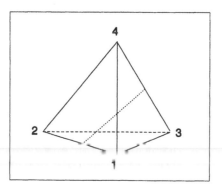

Figure 7.60 Tetrahedron for group operations.

S(3) is illustrated in Figures 7.58 and 7.59. The first columns show an equilateral triangle which can be transformed into itself by reflections along its symmetry axes or by rotation around its center. The respective operations are termed E (unity operation) A,B,C,D,F (set elements of S(3)). With these operations, the triplet of numbers 1,2,3 is permuted. The elements E,D,F represent a subset. One of the three numbers is assigned to each pair of exchange units connected by solution circuits (Figure 7.56). When operations A,B,C,D,F and E are applied, the six classes of the configurations of Figure 7.1u to z (set S_0) are obtained. In a similar way, the six classes of the sets S_1, S_2 and S_3 can be generated. In this case, one of the objects permutated in its relative location contains three exchange units.

Also, the permutations of the arrangements of pairs connected by horizontal vapor lines, Figure 7.57, yield the configurations of Figure 7.1u to z, representing the set S_0. In addition, the sets S_4, S_5, and S_6 are obtained when the combination of three exchange units from Figure 7.57 labeled "4" is included.

The transitions between the 24 configurations of the sets S_0, S_1, S_2, S_4 and S_0, S_4, S_5, S_6 can be combined into a common set which is isomorph to the symmetry operations of an equilateral tetrahedron (Figure 7.60). A rotation around the dotted axis exchanges corner 3 with corner 4 in the base plane. This corresponds to the transition from S_0 to S_1 in Figure 7.58 or to the transition from S_0 to S_5 in Figure 7.59. By applying similar operations, the two remaining sets are obtained, with E being replaced by a rotation.

The configurations of Figure 7.1s to t can be generated through combinations of elements from Figures 7.56 and 7.57 when two or three similar elements are used.

REFERENCES

1. Niebergall, W. (1981): Sorptions Kältemaschine Handbook der Kütetechmik Bd 7 Springer Verlag, Berlin.

2. Altenkirch, E. (1912): Reversible Absorptionsmaschinen. In: Zeitschrift für die gesamte Kälteindustrie, Jg. 20, Heft 1, S. 2-9; Heft 6, S. 114-119; Heft 8, S. 150-161; Jg. 21 (1914), Heft 1, S. 7-14; Heft 2, S. 21-24.

3. Alefeld, G. (1982a): Regeln für der Entwurf von mehrstufigen Absorptionswmepumpen. In: Brennstoff-Wärme-Kraft, Bd. 34, Nr. 2, S. 64-73.

4. Alefeld, F. (1983): Wärmeumwandlungssysteme, Skriptum zur Vorlesung am Institut für Technische Physik E 19, Physik-Department der Technischen Universität München.

5. Alefeld, G. (1982b): Kompression-und Expansionsmaschinen in Verbindung mit Absorberkreisläufen. In: Brennstoff-Wärme-Kraft, Bd. 34, Nr. 3, S. 142.

6. Alefeld. G. (1981): Wärmepumpentransformation, Brennstoff-Wärme-Kraft, vol 33, pg. 486-490.

CHAPTER EIGHT

Three and Multistage Absorption Configurations

Rule 4 in Chapter Five describes how multistage configurations can be obtained based on the two-stage configurations that were discussed in Chapter Seven. In the following, some configurations of special interest are discussed.

8.1 OPERATING MODES OF THREE-STAGE UNITS WITH FIVE TEMPERATURE LEVELS

Figures 8.1 and 8.2 show configurations with eight exchange units obtained from three basic cycles by merging twice two exchange units. The amounts of heat exchanged in the exchange units on the temperature levels T_0 through T_4 are the following for Figure 8.1a:

$$Q_A = n \qquad Q_D = -n \qquad Q_4 = n$$

$$Q_B = m - n \qquad Q_E = n - m \qquad Q_3 = m - 2n$$

$$Q_C = 1 - m \qquad Q_F = m - 1 \qquad Q_2 = n + 1 - 2m$$

$$Q_G = -1 \qquad Q_H = 1 \qquad Q_1 = m - 21$$

$$Q_0 = 1$$

In Figure 8.3a and b, the operating ranges are illustrated for the two conditions $n > 0$ and $n < 0$. These are limited by the cases $Q_3 = 0$, $Q_2 = 0$, and $Q_1 = 0$, which can be fulfilled by internal heat exchange on the temperature level indicated by the subscript 3, 2, or 1. Chapter One shows that there are 22 different reversible operating ranges in the case of 5 temperature levels. These are summarized in Figure 8.4. The configuration of Figure 8.1a can be operated in all 22 operating ranges, as indicated in Figures 8.3a and b. An equal number of operating ranges is obtained for the configurations of Figures 8.1b, c, and d and of Figure 8.2. Figures 8.3a and b have to be applied analogously. The table in Figure 8.4 is arranged so that two sequential columns differ by one sign only. It is the sign of that heat which is zero at the boundary between the two operating ranges. By crossing boundaries between two ranges in one corner, two signs change simultaneously. The operating ranges 8 and 19 are bordering on seven other operating ranges. The boundaries of the ranges and especially the corners represent modes of operation that are particularly interesting. The intersections $Q_3 = 0$ and $Q_2 = 0$ [i.e., $m = 2n$ and $m = (n + 1/2)$ for Figure 8.1a] represent heat pumps with cooling COPs of $\xi_c = 3$ for $n > 0$, (triple-effect units) and heat transformers of huge temperature lift for $n < 0$. The intersections $Q_2 = 0$ and $Q_1 = 0$ [i.e., $m = 2$ 1 and $m = (n + 1)/2$ for Figure 8.1a] represent refrigeration units of $\xi_c = 1/3$ for $n > 0$. A temperature T_0 is reached which, on the reciprocal temperature scale, is three times the temperature difference between driving heat and waste heat below the waste heat temperature level. For $n < 0$, this operating mode represents a heat transformer with a COP of $\xi_h = 3/4$ with a small temperature lift. The same method applied to other intersections produces analogous results. This holds also for the cases of $m = 0$ and $n = 0$.

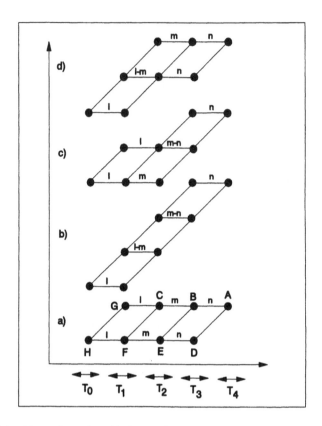

Figure 8.1   Absorption units with three independent cycles and eight exchange units.

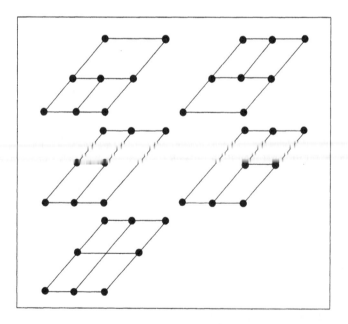

Figure 8.2   Absorption units with three independent cycles and eight exchange units.

The configuration of Figure 8.1b with internal heat exchange in the highest and lowest cycle is an extension of the "double-effect" machine for LiBr/H_2O. This unit may be operated as a heating unit similar to the one described in Figure 7.40, (Chapter Seven), yet obtaining an even higher COP.

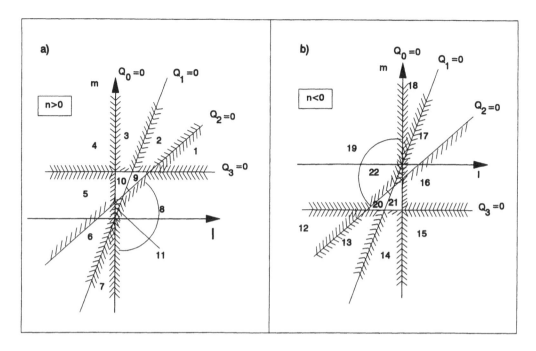

Figure 8.3 Operating ranges of the configurations in Figure 8.1a.

	1	2	3	4	5	6	7	8	9	10	11	12	13	14	15	16	17	18	29	20	21	22
Q_4	+	+	+	+	+	+	+	+	+	+	+	−	−	−	−	−	−	−	−	−	−	−
Q_3	+	+	+	+	−	−	−	−	−	−	−	−	−	−	−	+	+	+	+	+	+	+
Q_2	+	−	−	−	−	+	+	+	−	−	+	−	+	+	+	+	−	−	−	+	+	+
Q_1	−	−	+	+	+	+	−	−	−	+	+	+	+	−	−	−	−	+	+	+	−	−
Q_0	+	+	+	−	−	−	−	+	+	+	+	−	−	−	+	+	+	+	−	−	−	−

Figure 8.4 Sign matrix for the operating ranges 1–22 for the configurations of Figures 8.1 and 8.2.

8.2 THREE-STAGE ABSORPTION CONFIGURATIONS

The multiplicity of three-stage configurations is now used as an example to demonstrate the diversity of multistage configurations.

A. When expanding the configuration of Figure 7.1a to r (Chapter Seven) by adding one basic cycle which has one exchange unit in common with the original configuration, but which is not a common unit to the original two cycles, the following characteristics are obtained:

10 exchange units, 3 working fluid streams

9 exchange units, 2 working fluid streams (Rule 3a)

8 exchange units, 1 working fluid stream (Rule 3a)

B. When expanding the configurations of Figure 7.1a to r by one basic cycle which has one exchange unit in common with the exchange units in common with the original two basic cycles, configurations with the following characteristics are obtained:

10 exchange units, 3 working fluid streams

9 exchange units, 2 working fluid streams (Rule 3a)

Rule 3a cannot be applied any further to these configurations. Any further application of Rule 3a would result in two-stage configurations. However, Rule 3b can be used further. A similar situation occurs when Rule 3a is applied to the configurations of Figure 7.1s and t.

C. When expanding configurations of Figure 7.1a to r by one basic cycle so that it has two exchange units in common with the original configuration, configurations with the following characteristics are obtained:

9 exchange units, 3 working fluid streams

8 exchange units, 2 working fluid streams (Rule 3a)

A subset (and only a subset) of the configurations containing nine exchange units is obtained by extending the configurations from Figure 7.1s and t by one basic cycle which has one exchange unit in common with the original configuration. These configurations of eight exchange units can also be obtained by extending the configurations of Figure 7.1u to z by a basic cycle which has two exchange units in common with the original one.

These configurations are illustrated in Figures 8.5 to 8.10. However, care should be taken when using this shortcut that all classes are obtained. The application of the shortcut implies that Rule 4 and Rule 3a are commutative. This is not always valid. Furthermore, when adding a basic cycle to a configuration, care has to be taken to generate only one additional loop. The configurations of Figure 7.1a,b,h,i,m,n,q,r can all be expanded by adding to it one basic cycle which closes two additional loops. This creates a four-stage machine of nine exchange units and four independent fluid streams as shown in Figure 8.13. Three-stage configurations (quadruple-effect) of high COP are illustrated in Figure 7.26, Chapter Seven.

D. When expanding the configurations of Figure 7.1a to r by another basic cycle which has three exchange units in common with the original one, configurations of the following characteristics are obtained: 8 exchange units, 3 working fluid streams. If Rule 3a were applied, configurations of fewer stages would result.

E. When expanding the configurations of Figure 7.1s and t by another basic cycle so that the original and the new cycle have two exchange units in common, the configurations mentioned in D, above, plus two additional ones (Figures 8.1a and b) are obtained with the following characteristics: 8 exchange units, 3 working fluid streams.

All configurations described in E, above, are displayed in Figure 8.11 (extension of Figure 7.1t) and Figure 8.12 (extension of Figure 7.1s). The number of topologically different classes in Figures 8.11 and 8.12 totals 11, several of which have been shown in Figures 8.1 and 8.2.

One recognizes that combining basic cycles twice, so that they have two exchange units in common, results in two additional configurations. These cannot be obtained by combining two basic cycles with one common exchange unit in such a manner to a third one, so that the third cycle has three exchange units in common with the first two.

Summary

The following characteristics are found for three-stage absorption configurations if they are classified according to the number of independent fluid streams:

Figure 8.5 Configurations with eight exchange units.

Figure 8.6 Configurations with eight exchange units.

Figure 8.7 Configurations with eight exchange units.

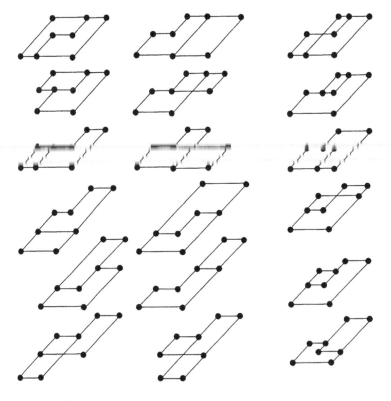

Figure 8.8 Configurations with eight exchange units.

Figure 8.9 Configurations with eight exchange units.

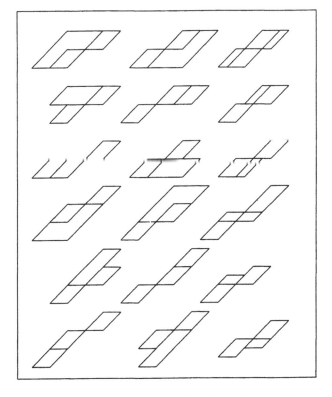

Figure 8.10 Configurations with eight exchange units.

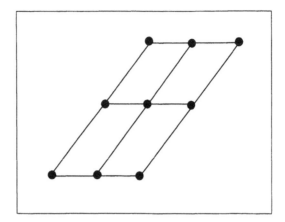

Figure 8.11 Configuration with nine exchange units.

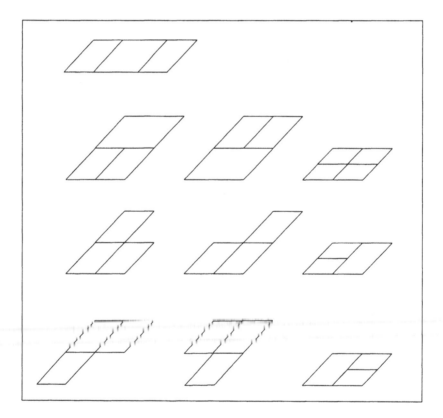

Figure 8.12 Configurations with eight exchange units.

3 working fluid streams: 10, 9, or 8 exchange units

2 working fluid streams: 9 or 8 exchange units

1 working fluid stream: 8 exchange units

8.3 MULTISTAGE ABSORPTION CONFIGURATIONS

The most general four-stage absorption configuration contains 13 exchange units and 4 independent working fluid streams. By applying Rule 3a three times, configurations of ten exchange units and one working fluid stream are obtained.

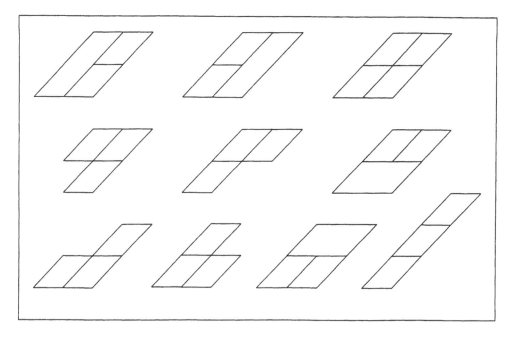

Figure 8.13 Configurations with eight exchange units.

The maximum number of exchange units E_{max} is obtained by the following equation where s represents the number of stages:

$$E_{max} = 3s + 1$$

The minimum number of exchange units obtained after applying Rule 3a (s – 1) times is given by:

$$E_{min} = 2s + 2$$

It should be pointed out that the absolute minimum of exchange units for a given number of stages can be lower, because the added basic cycle may have up to four exchange units in common with the original configuration. With s = 4, one additional configuration (Figure 8.13) containing nine exchange units is obtained with four independent working fluid streams. For s ≥ 3, the maximum number of exchange units for a given stage can be equal to or larger than the minimum number of the next stage. For example, ten exchange units can be combined to three-stage and four-stage configurations.

Figure 8.14a and b shows how three-stage configurations can be expanded to four-stage configurations by adding one basic cycle which has four exchange units in common with the three-stage configuration. This expansion consists of adding one new connection shown as a dashed line.

The number of classes K(s) of configurations increases exponentially as is mostly the case with permutations. For s = 4 more than ten thousand configurations can be identified. It is emphasized again, that the discussion in this chapter has been limited to those cases in which all cycles are heat and fluid coupled. It is obvious that using the same combination methods all cascading (only heat coupled) cycles can be identified too. Also those combinations of cascades can be found in which individual cascades are composed of multistage cycles.

8.4 EVALUATION OF THE COP OF MULTISTAGE ABSORPTION CYCLES

A method is presented to calculate the COP of advanced absorption cycles very quickly if the efficiencies of single-stage absorption heat pumps or absorption heat transformers are known. The method provides detailed results for the heat input and output of individual exchange

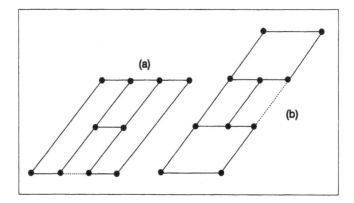

Figure 8.14 Extension of three-stage configurations to four-stage configurations by adding a basic cycle which has four exchange units in common with the original configuration.

units. These values can be used for the comparison of cycles. Examples are given for absorption heat pumps, refrigerators, heat transformers, and heat pump transformers for the working fluid pairs water/lithiumbromide and ammonia/water.

8.4.1 Basic Single-Stage Cycle

The discussion begins with the single-stage cycles to demonstrate the concept.

Single-Stage Refrigerator

Equations are derived for the quantities of heat to be exchanged at the four exchange units of a single-stage absorption refrigerator. The evaluation is based on 1 kg of refrigerant circulating in the cycle. Figure 8.15 shows schematic diagrams of the single-stage refrigerator and heat transformer:

$$q_E = r_E \qquad\qquad - c(T_1 - T_0) \;\; - v\Delta p$$

$$q_c = r_c \qquad\qquad + c_p(T_3 - T_1)$$

$$q_G = (r+1)_G + q_{HX} + c(T_3 - T_2)$$

$$q_A = (r+1)_A + q_{HX} - c_p(T_2 - T_0) \qquad + (f-1)v_1\Delta_p$$

$$w_s = \qquad\qquad\qquad\qquad\qquad f v_1 \Delta p$$

$$\qquad\qquad A \quad\;\; B \qquad\;\; C \qquad D \qquad\qquad\qquad (8.1)$$

v	=	Specific volume, liquid refrigerant
v_1	=	Specific volume, solution with low refrigerant content
v_h	=	Specific volume, solution with high refrigerant content
r	=	Latent heat
1	=	Heat of mixing
w_s	=	Pump work for solution
c	=	Isobaric specific heat capacity of liquid refrigerant
c_p	=	Isobaric specific heat capacity of gaseous refrigerant

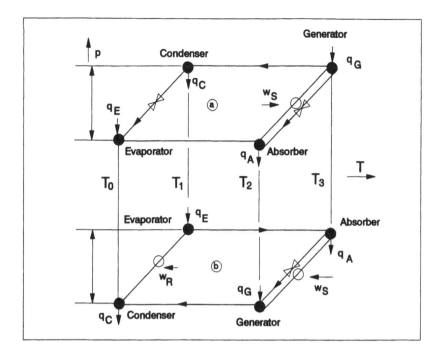

Figure 8.15 Schematic diagrams of single-stage absorption devices.

c_l = Isobaric specific heat capacity of solution with low refrigerant content

c_h = Isobaric specific heat capacity of solution with high refrigerant content

Four terms, A, B, C, and D, of different physical origin are contributing to the exchanged heat quantities.

The first term, A, describes the latent heat of phase change. In the case of the evaporator and condenser, it is the heat of evaporation, r, at the corresponding temperature. In the generator and absorber, this heat is increased by the differential heat of solution, l.

The second term, B, is the amount of heat, q_{HX}, which flows from the generator to the absorber as specific heat of the poor solution due to the finite surfaces of the solution heat exchanger. In a heat exchanger with infinitely large surface, all of the specific heat, $(f - 1)c_l(T_3 - T_2)$, of the poor solution is used to heat the rich solution. In a real heat exchanger, only a fraction of this heat, e_{HX}, can be used; e_{HX} is called heat-exchanger effectiveness. So the heat which is not recovered amounts to:

$$q_{HX} = (1 - e_{HX})(f - 1)c_l(T_3 - T_2) \tag{8.2}$$

The third term, C, describes sensible heat, which has to be supplied or removed to heat or cool a flow of liquid (heat capacity c) or vapor refrigerant (heat capacity c_p). The sensible heat term in the equation for the generator heat, q_G, shows up due to the mismatch between the mass flow and heat capacity of rich and poor solution in the solution circuit. To arrive at this equation, the assumption of a linear change in the specific heat capacity of the solution with regard to composition is applied:

$$c_h f - c_l(f - 1) = c \tag{8.3}$$

where c_l and c_h are the specific heat capacities of the solution with low or high content of refrigerant; f is the circulation ratio, defined as mass of rich solution circulating in the solution circuit per mass of refrigerant being desorbed.

The term D stands for the mechanical work input to the cycle and the dependence of enthalpy on pressure. To account for irreversibilities of the pump, an efficiency could be introduced into the equations. The mechanical work, which is converted to heat due to irreversibilities, mainly increases the heat load of the absorber, but also reduces slightly the input into the generator. For the present purpose, these effects are not important and, therefore, only the isentropic pump work $v_h \Delta p$ is considered. This pumping power is converted to heat by irreversible throttling in the pipes to the evaporator and absorber, $v \Delta p$ and $v_l(f - 1)\Delta p$, respectively.

For the specific volumes, v, of liquid refrigerant, v_h of rich and v_l of poor solution, an analogous mixing rule, as for the heat capacities, has been assumed to hold:

$$v_h f - v_l (f - 1) = v \tag{8.4}$$

Now term A is analyzed in more detail. The heat of evaporation, r, is changing with temperature, so $r_E = r_C + r$ for $T_C = T_E + T$. The temperature dependence of the heat of evaporation can be deduced by the enthalpy balance of a reversible Rankine Cycle:

$$r_E = r_C + v\Delta p + (c - c_p)(T_1 - T_0) \tag{8.5a}$$

This is part of "Planck's formula".

In the desorber and the absorber, the latent heat is increased by the heat of solution, l. By calculating a Rankine-Sorption Process in a manner analogous to the standard Rankine Cycle above we find, using Equations 8.3 and 8.4:

$$(r + l)_A = (r + l)_G + v\Delta p + (c - c_p)(T_3 - T_2) \tag{8.5b}$$

Inserting Equations 8.5a, 8.5b and 8.4 into Equation 8.1, the following is obtained:

$$q_E = r_c \qquad \qquad -c_p(T_1 - T_0)$$

$$q_c = r_c \qquad \qquad +c_p(T_3 - T_1)$$

$$q_G = (r + l)_G + q_{HX} \qquad +c(T_3 - T_2)$$

$$q_A = (r + l)_G + q_{HX} \qquad -c_p(T_3 - T_0) + c(T_3 - T_2) + fv_h\Delta p$$

$$w_s = \qquad \qquad fv_h\Delta p \tag{8.6}$$

We define the COP of the single-stage refrigerator as heat ratio, η, refrigeration effect divided by desorber heat input:

$$\eta = COP = q_E / q_G \tag{8.7}$$

Using Equations 8.6 and 8.2, the single-stage efficiency finally is calculated to:

$$\eta = \frac{r_C - c_p(T_1 - T_0)}{(r + l)_G + [c + c_l(f - 1)(1 - e_{HX})](T_3 - T_2)} \tag{8.8}$$

Table 8.1 Efficiency η and η_T and Asymmetry and Pump Factors α_i with Data Used for Calculation

	LiBr/Water		Ammonia/Water	
	Heat pump	**Transformer**	**Heat pump**	**Transformer**
$T_0(°C)$	5	35	−10	35
$T_1(°C)$	35	70	35	70
$T_2(°C)$	35	75	35	75
$T_3(°C)$	75	110	95	110
$r_C(kJ/kg)$	2417	2417	1123	1123
$(r+1)/r$	1.10	1.11	1.15	1.10
$c(kJ/kgK)$	4.2	4.2	4.7	5.0
$c_p(kJ/kgK)$	1.9	1.9	2.1	2.1
$c_1(kJ/kgK)$	2.0	2.0	4.7	4.7
$v_h(dm^3/kg)$	0.6	0.6	1.3	1.3
$v(dm^3/kg)$	1.0	1.0	1.7	1.7
$\Delta P(bar)$	0.046	0.245	10.6	19.6
$f(kg/kg)$	19	29	24	17
e_{HX}	0.95	0.95	0.95	0.95
η,η_T	0.79	0.48	0.57	0.40
a,a_T	0.056	−0.004	0.21	−0.006
a_{TPS}		0.0002		0.034
a_p	0.0001		0.032	
a_{TPR}		0.00001		0.003

To estimate the COP of multistage cycles, the expression in Equation 8.8 is used as a parameter. For this reason, two dimensionless factors are introduced into the equations.

The "asymmetry factor", a, describes the relative difference between condenser and evaporator heat and is given by:

$$a = \frac{q_C - q_E}{q_E} = \frac{c_p(T_3 - T_0)}{r_C - c_p(T_1 - T_0)} \tag{8.9a}$$

The factor a_p is the ratio between pump work and evaporator heat:

$$a_p = \frac{w_s}{q_E} = \frac{fv_h \Delta p}{r_C - c_p(T_1 - T_0)} \tag{8.9b}$$

a and a_p are both small as compared to 1 (see Table 8.1). In contrast to the efficiency, these factors are not influenced by the heat exchanger loss, q_{HX}.

The set of Equations 8.6 is divided by q_G, and the definitions of Equations 8.7, 8.9a and 8.9b are used. So by simple algebraic operations, the following reduced heat loads are obtained:

$$q_E' = \eta$$

$$q_C' = \eta + \eta a$$

$$q_G' = 1$$

$$q_A' = 1 - \eta(a - a_p)$$

$$w_s' = \eta a_p \tag{8.10}$$

Single-Stage Heat Transformer
The thermal loads at the individual exchange units can be expressed as follows:

$$q_E = r_E \qquad\qquad +c(T_1 - T_0)$$

$$q_c = r_c \qquad\qquad +c_p(T_2 - T_0)$$

$$q_G = (r+1)_G - q_{HX} \quad -c(T_2 - T_2) \qquad -fv_h \eta p$$

$$q_A = (r+1)_A - q_{HX} \quad -c_p(T_3 - T_1)$$

$$w_s = \qquad\qquad\qquad\qquad (f-1)v_l \Delta p$$

$$w_r = \qquad\qquad\qquad\qquad v \Delta p$$

$$\qquad A \qquad\quad B \qquad\qquad C \qquad\qquad D \qquad\qquad (8.11)$$

For the individual terms of Equations 8.9a and 8.9b, the same explanation holds as for the set of Equations 8.1. The isentropic work of the solution pump, $w_s = (f-1)v_l p$, as well as that of the refrigerant pump, $w_R = vp$, is converted to heat, fv_{hp}, in the throttling process and becomes available in the desorber.

The condenser pressure in the heat transformer is lower than the evaporator pressure. Therefore, we find for the temperature dependence of the latent heat:

$$r_E = r_C - v\Delta p \qquad\qquad (c - c_p)(T_1 - T_0)$$

$$(r+1)_A = (r+1)_G - v\Delta p - \qquad (c - c_p)(T_3 - T_2) \qquad\qquad (8.12)$$

$$q_E = r_C \qquad\qquad +c_p(T_1 - T_0) \qquad\qquad\qquad -v\Delta p$$

$$q_c = r_C \qquad\qquad +c_p(T_2 - T_0)$$

$$q_G = (r+1)_G - q_{HX} \quad -c(T_3 - T_2) \qquad\qquad\qquad -fv_h \Delta p$$

$$q_A = (r+1)_G - q_{HX} \quad -c_p(T_2 - T_1) \quad -c(T_3 - T_2) \quad -v\Delta p$$

$$w_s = \qquad\qquad\qquad\qquad\qquad\qquad\qquad (f-1)v_l \Delta p$$

$$w_r = \qquad\qquad\qquad\qquad\qquad\qquad\qquad v\Delta p \qquad\qquad (8.13)$$

The COP of the single-stage heat transformer is defined as the heat ratio, η_T, between the heat output of the absorber and the heat input to the desorber and the evaporator:

$$\eta_T = COP_T = q_A / (q_G + q_E) \qquad\qquad (8.14)$$

The single-stage efficiency of the heat transformer results from Equations 8.13 and 8.2:

$$\eta_T = \frac{(r+1)_G - c_p(T_2 - T_1) - [c + c_l(f-1)(1 - e_{HX})](T_3 - T_2) - v\Delta p}{r_C + (r+1)_G + c_p(T_1 - T_0) - [c + c_l(f-1)(1 - e_{HX})](T_3 - T_2) - (v + fv_h)\Delta p} \qquad (8.15)$$

The asymmetry factor, a_T, describing the relative difference in heat load of evaporator and condenser is now given by:

$$a_T = \frac{q_C - q_E}{q_c} = \frac{c_p(T_2 - T_1) + v\Delta p}{r_C + c_p(T_2 - T_0)} \tag{8.16a}$$

The factor a_{TPS}, describing the mechanical work of the solution pump is defined as:

$$a_{TPS} = \frac{W_s}{q_c} = \frac{(f-1)v_1\Delta p}{r_C + c_p(T_1 - T_0)} \tag{8.16b}$$

The factor a_{TPR}, describing the mechanical work of the refrigerant pump is defined as:

$$a_{TPR} = \frac{W_R}{q_C} = \frac{v\Delta p}{r_C + c_p(T_2 - T_0)} \tag{8.16c}$$

a_T, a_{TPS}, and a_{TPR} are again small compared to 1. The set of Equations 8.13 is now divided by $(q_G + q_E)$ and the definitions of Equations 8.14, 8.16a, 8.16b, and 8.16c are used. The following reduced heat loads are obtained:

$$q'_E = (1 - \eta_T) - (1 - \eta_T)\frac{a_T - a_{TP}}{1 - a_{TP}}$$

$$q'_C = (1 - \eta_T) + (1 - \eta_T)\frac{a_{TP}}{1 - a_{TP}}$$

$$q'_G = \eta_T + (1 - \eta_T)\frac{a_T - a_{TP}}{1 - a_{TP}}$$

$$q'_A = \eta_T$$

$$w'_s = (1 - \eta_T)\frac{a_{TPS}}{1 - a_{TP}}$$

$$w'_R = (1 - \eta_T)\frac{a_{TPR}}{1 - a_{TP}} \tag{8.17}$$

In Table 8.1 the efficiencies, η_i, factors, a_i, and the thermodynamic data required are listed for typical heat pump and heat transformer cycles for the two working pairs $H_2O/LIBr$ and NH_3/H_2O, respectively. Note that the temperature lift of the heat pump with NH_3/H_2O is larger than with $H_2O/LiBr$. This is one reason for the considerably lower efficiency, η, of the former.

The significance of the factors a_i is as follows. First the fluid pair, $H_2O/LiBr$, is considered. The pump work is small. The asymmetry factor, a_T, contains the difference, $(T_1 - T_2)$, between the evaporator and generator temperature, which is usually small. The validity of the estimation of a_T can be verified by a calculation of an exact enthalpy balance of a heat transformer with $T_1 > T_2$ and another with $T_1 < T_2$. It is found that, in the first case, the heat load of the evaporator is larger than that of the condenser, while in the second case the reverse is true. This change in sign of a_T is predicted by Equation 8.16a. For $T_1 \approx T_2$, $a_T \approx 0$. The asymmetry factor a of the heat pump is, for most cases, smaller than 0.1.

In the case of NH_3/H_2O, all the factors are considerably larger than in the case for $H_2O/LiBr$. Especially for the heat pump, the asymmetry factor, a, may amount to even more than 20%. However, it may be reduced by rectification of the vapor flowing from the desorber to the condenser and preheating the vapor flowing from the evaporator to the absorber by subcooling liquid condensate. This is usually done with this working pair.

Ignoring the pump work and using the definitions for the efficiencies and the asymmetry factors, the four heat quantities of the single-stage heat pump and heat transformer can be written as follows:

heat pump:

$$q_G = 1$$
$$q_E = \eta$$
$$q_c = \eta + \eta a$$
$$q_A = 1 - \eta a$$

heat transformers:

$$q_A = \eta_T$$
$$q_c = 1 - \eta_T$$
$$q_E = 1 - \eta_T \quad -a_T(1 - \eta_T)$$
$$q_G = \eta_T \quad +a_T(1 - \eta_T)$$

These relationships follow directly from definitions and the application of the First Law, without knowing the precise values for a and a_T.

The intention is not to give an exact description of the single-stage cycle, but to provide means for a fast estimation of multistage cycle efficiency with an accuracy of about 10%. Therefore, the asymmetry factors in the following calculations are neglected. This approximation is nearly exact for heat transformers and very good for heat pumps with $H_2O/LiBr$ as a working pair. It is less appropriate in the case of NH_3/H_2O. It still depends on the design of the multistage cycle itself, how large the error due to the approximation finally is. There are some cycles (R5 and R6 in Table 8.5, for instance) where the asymmetry factors would not appear in the final results, even if they were included in the calculation.

If the asymmetry factors are neglected, the heat loads, q', from Equations 8.10 and 8.17 are simplified, as shown in the first column, a, of Tables 8.2 and 8.3. The values are based on one heat set to unity. In the second and third columns of these tables, the heat loads are shown with the heat load of the other exchange units being set to unity. In Table 8.4, single-stage efficiencies are listed as they are reached in commercial machinery with the working pairs $H_2O/LiBr$ and NH_3/H_2O, respectively. These values will be used in the following calculations.

8.4.2 COPs of Multistage Cycles Using the Principle of Superposition
Rules for Superposition
Basic rules for the synthesis of multistage cycles have been presented in Chapter Five. The essentials of those rules are as follows: Every multistage cycle can be decomposed into a cascade of single-stage cycles called basic cycles. The basic cycles of the cascade are coupled by heat transfer. This decomposition into cascades of elementary cycles may require more exchange units than actually exist in the multistage cycle. Nevertheless, the heat and mass flow rates in a cascade are not very different from those in an integrated multistage system, and the heat ratios and thus the COPs of the cascades are nearly the same as for the integrated systems as well. Therefore, the heat ratios and COP of a multistage cycle can be estimated by calculating the heat ratios of the respective cascade cycle using the heat ratios of the elementary cycles. The best approximation is achieved if the following rules are observed (see Figure 8.15):

Table 8.2 Simplified Heat Loads of Single-Stage Refrigerator or Heat Pump, Normalized to: (a) Generator or Absorber heat; (b) Evaporator or Condenser Heat; (c) Reject Heat of Absorber Plus Condenser

	a	b	c
Evaporator	η	1	$\eta/(1+\eta)$
Condenser	η	1	1
Absorber	1	$1/\eta$	1
Generator	1	$1/\eta$	$1(/1+\eta)$

Table 8.3 Simplified Heat Loads of Single-Stage Heat Transformer, Normalized to: (a) Heat Input to Evaporator Plus Generator; (b) Condenser or Evaporator Heat; (c) Generator or Absorber Heat

	a	b	c
Condenser	$1 - \eta_T$	1	$(1 - \eta_T)/\eta_T$
Evaporator	1	1	$(1 - \eta_T)/\eta_T$
Generator	1	$\eta_T/(1 - n_T)$	1
Absorber	$1 + \eta_T$	$\eta_T/(1 - n_T)$	1

Table 8.4 Single-Stage Efficiencies

Working fluid pair	H_2O/LiBr	NH_3/H_2O
Refrigerator η	0.72	0.55
Heat transformer ηT	0.47	0.42

1. Solution circuits of the multistage cycle with a clockwise flow direction are to be represented in the cascade by a heat pump cycle.

2. Solution circuits of the multistage cycle with a counterclockwise flow direction are to be represented in the cascade by a heat transformer cycle.

These rules for decomposition and superposition ensure that the major irreversibilities of the multi-stage cycle are properly taken into account. To illustrate the method, some examples will be discussed.

Double-Effect Chiller

This cycle is shown in Figure 8.16 (cycle C). There are three ways of composing the cycle. In Figure 8.16a it is composed of a heat pump, A, and a heat transformer, B. In a cascade of this kind, the solution circuit of B counteracts that of A, but their irreversibilities add up. This is not the case in a multistage system where the solution circuits partly cancel. So the solution flow and consequently the irreversibility originating from the solution circuit are smaller in the integrated cycle than in the cascade. The double-effect chiller has only solution circuits with the solution flowing clockwise. So, following the above formulated rule, we have to use only heat pump cycles to compose the machine. This is shown in Figures 8.16b and c. In these two cascades, no additional irreversibilities occur due to counteracting solution circuits. So their heat ratios will be closer to that of the integrated multistage system than the ratios resulting from the cascade in Figure 8.16a.

To calculate the COP of the cascades in Figures 8.16b and c, the heat load of the evaporator and condenser of cycle A are chosen to be unity in both cases. In Figure 8.16b, the heat of condensation of cycle A is supplied to the desorber of cycle B. The heat load of the desorber of cycle B is unity as well. The heat loads of the other exchange units are found according to

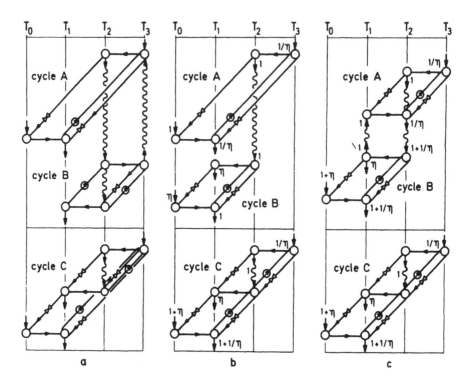

Figure 8.16 Synthesis of a double-effect chiller. (a) A, heat pump; B, heat transformer; C, superposition forbidden; (b) and (c), A and B heat pumps; C, superposition of A and B.

Table 8.2. In Figure8.16c, first the heat load of the absorber of cycle A has to be found which is 1/ according to Table 8.2. The heat load of the desorber and absorber of cycle B is $1 + 1/\eta$, which yields $\eta(1 + 1/\eta) = \eta + 1$ for the evaporator and condenser load. From the condenser of cycle B one unit of heat has to be supplied back to the evaporator of cycle A, which leaves $\eta(1 + 1/\eta) - 1 = \eta$ to be rejected to the heat sink.

Both methods in Figures 8.16b and c yield the same result for the cooling COP_R of the double-effect chiller:

$$COP_R = \eta(1+\eta) = \eta + \eta^2 \tag{8.18}$$

For $\eta = 1$, the COP would be doubled. For a real machine, the single stage efficiency is not doubled, but multiplied by $1 + \eta$, with about 1.7 for $H_2O/LiBr$. With a single-stage efficiency for $H_2O/LiBr$ of $\eta = 0.72$, a COP_R of 1.24 is obtained. This is, in fact, the efficiency (not including the burner) of commercially available double-effect chillers. The heat ratios of multistage heat transformers can be found fairly analogously.

Heat Pump Transformer

These are advanced cycles with four external temperature levels, with the following signs required for the heat fluxes: $(+ - + -)$ or $(- + - +)$; see Chapter One. Examples have been discussed in Chapter 7.2, and in Figure 8.17 an additional one is shown. Figure 8.18 demonstrates how the heat pump transformer can be applied as a heat supply system to a separation process. A heat pump transformer is basically a combination of a heat pump and a heat transformer. The amount of heat which is not delivered to the process by the heat transformer section is supplied by the heat pump section sitting on top of the transformer section. At the highest temperature, T_3, driving heat Q_3 is supplied to the heat pump transformer. At T_2, heat Q_2 is transferred from the heat pump transformer to the separation process. The section Q_{loss}

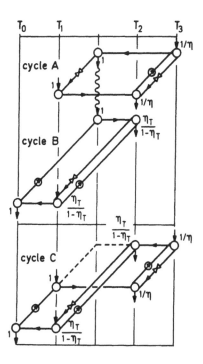

Figure 8.17 Synthesis of heat pump transformer.

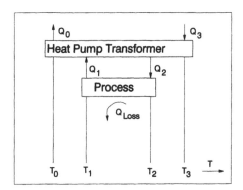

Figure 8.18 Heat pump transformer as a heat supply system for a separation process.

of this heat is lost in the separation process; the rest, Q_1, is recovered by the heat pump transformer at T_1 and upgraded to T_2. At the lowest temperature T_0, heat Q_0 is rejected to the ambient by the heat pump transformer.

The cycle C, displayed in Figure 8.17, can be decomposed into a heat pump cycle, A, and a heat transformer cycle, B. The heat loads of evaporator and condenser of both cycles are chosen to be unity. Therefore, the condenser of cycle A and the evaporator of cycle B cancel, as is shown in Figure 8.17c. The COP_{HPT} of the cycle is defined as output Q_2 to the process at temperature T_2, divided by the input of driving heat Q_3 to the absorption system at T_3:

$$COP_{HPT} = Q_2/Q_3 = 1 + \eta\eta_T/(1-\eta_T) \tag{8.19}$$

which is 1.64 with the single-stage efficiencies for $H_2O/LiBr$ from Table 8.4. The ratio of heat input to the process, Q_2, and heat recovered from the process, Q_1, is:

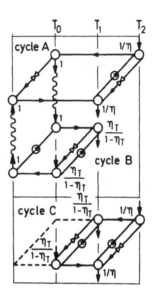

Figure 8.19 Synthesis of a resorption cycle; A, heat pump; B heat transformer; C, superposition of A and B.

$$Q_2/Q_1 = \eta_T + (1-\eta_T)/\eta \qquad (8.20)$$

which is 1.17 for $H_2O/LiBr$. So the heat supply system allows for a 15% heat loss of the process.

Resorption Cycle

This single-stage cycle, C, in Figure 8.19, having two solution circuits is different from a common single-stage cycle. The cycle is decomposed into one heat pump (A) and one heat transformer (B) to incorporate the two solution circuits into the resorption cycle. The heat loads in the evaporator and condenser of both cycles must be equal and have been chosen to be unity. The heat loads of the other exchange units are found using Tables 8.2 and 8.3. By combining cycle A and cycle B we obtain the resorption cycle, C. The COP_R for cooling of a resorption cycle is found to be

$$COP_R = \eta\eta_T/(1-\eta_T) \qquad (8.21)$$

which is smaller than η, as long as η_T is smaller than 1/2. With the values of Table 8.3 for η and η_T, a $COP_R = 0.64$ is obtained for $H_2O/LiBr$ and $COP_R = 0.33$ for NH_3/H_2O. In this special case, the superposition is only strictly possible if the asymmetry factors can be neglected.

8.4.3 Results

Using the methods for cycle analysis described above, a number of advanced cycles will be compared based on COP and energy savings. For this purpose, a number e is defined that serves as a measure of the energy savings resulting from the application of an advanced cycle (heat input Q_{in}) compared to a reference system (heat input $Q_{in,ref}$).

$$e = (Q_{in,ref} - Q_{in})/Q_{in,ref} \qquad (8.22)$$

Refrigeration Cycles

As a reference system, we take a single-stage refrigerator with the cooling capacity Q_{use}. With $Q_{use}/Q_{in,ref} = \eta$ of the single-stage cycle and $Q_{use}/Q_{in} = COP_R$ of the advanced cycle, Equation 8.24 yields for the energy-saving factor e:

Table 8.5 Double and Triple-Effect Cycles Compared to a Single-Stage Chiller with Regard to COP_R and Relative Energy Savings, e

REFRIGERATOR	COP_R	COP_R H$_2$O/LiBr / COP_R NH$_3$/H$_2$O	e(%) H$_2$O/LiBr / e(%) NH$_3$/H$_2$O
R1	η	0.72	Reference
		0.55	Reference
R2	$\eta(1+\eta)$	1.24	42
		0.85	35
R3	2η	1.44	50
		1.10	50
R4	$\dfrac{\eta}{1-\eta_T}$	1.36	47
		0.95	42
R5	$\eta(2+\eta)$	1.96	63
		1.40	61
R6	$\eta\dfrac{2-\eta_T}{1-\eta_T}$	2.08	65
		1.50	63
R7	$\eta\dfrac{1+\eta\eta_T}{1-\eta_T}$	1.82	60
		1.17	53

$$e = 1 - \eta/COP_R \tag{8.23}$$

The results for several refrigeration cycles are listed in Table 8.5 with the formula for calculating the COP_R as well as the energy-saving factor e. In Figure 8.20, COPs of these cycles are plotted.

The COP_R can be raised considerably by changing from single-effect to double-effect and further to triple-effect cycles. Using a double-effect cycle, the energy saving will be in the range of 40 to 50%. With a triple-effect cycle, more than 60% of the input energy can be saved.

It is important to note the difference in COP between H$_2$O/LiBr and NH$_3$/H$_2$O: the best double-effect cycle, R3, which is well suited for use with NH$_3$/H$_2$O, may reach a COP_R of 1.10, which is below that of the double-effect machine R2 for H$_2$O/LiBr with COP_R = 1.24. This value can be reached with NH$_3$/H$_2$O only if the triple-effect cycle R6 is being used. Consequently, only working pairs with comparable high single-stage efficiency can compete with H$_2$O/LiBr in advanced cycles. Unfortunately, many attractive cycles, like R3 and R4 or R6, cannot be used with H$_2$O/LiBr due to its narrow solution field.

Heat Pump Cycles

Direct heating is used as a reference for the energy-saving factor e. $Q_{in,ref}$ being equal to the useful heat output, Q_{use}, we find from Equation 8.22:

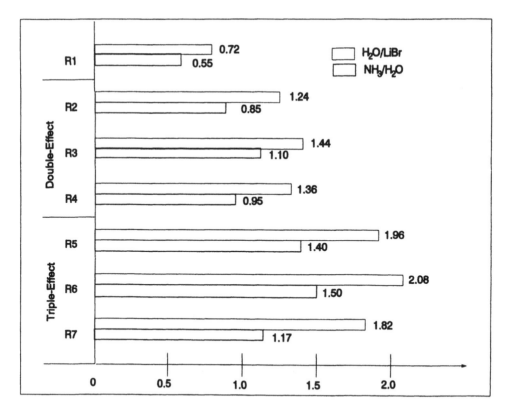

Figure 8.20 Coefficient of performance, COP$_r$ of several advanced refrigeration cycles.

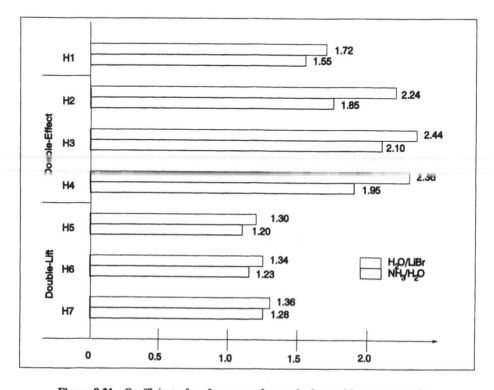

Figure 8.21 Coefficient of performance of several advanced heat pump cycles.

Table 8.6   Double-Effect and Double-Lift Cycles Compared to a Single-Stage Heat Pump and Direct Heating with Regard to COP_H and Relative Energy Saving, e

HEAT PUMP	COP_H	COP_H $H_2O/LiBr$	$e(\%)$ $H_2O/LiBr$
		COP_H NH_3/H_2O	$e(\%)$ NH_3/H_2O
H1	$1+\eta$	1.72	42
		1.55	35
H2	$1+\eta+\eta^2$	2.24	55
		1.85	46
H3	$1+2\eta$	2.44	59
		2.10	52
H4	$\dfrac{1+\eta-\eta_T}{1-\eta_T}$	2.36	58
		1.95	49
H5	$\dfrac{1+\eta+\eta^2}{1+\eta}$	1.30	23
		1.20	17
H6	$1+\eta\eta_T$	1.34	25
		1.23	19
H7	$1+\eta/2$	1.36	26
		1.28	21

$$e = 1 - 1/COP_H \tag{8.24}$$

for the energy-saving factor, e. This is shown in Table 8.6 together with COP_H. In Figure 8.21, the COPs are compared graphically. Cycles H1 to H4 correspond to R1 to R4 in Table 8.5, but the energy-saving factors are different because of the different reference system.

The best system for the double-effect cycles is again cycle H3. The energy savings of the three cycles, however, especially in the case of $H_2O/LiBr$, are not very different. Cycle H4 might be preferable because its investment cost is probably lowest.

The last three cycles in Table 8.6 and Fig 8.17 yield a high temperature lift. This is the reason for their relatively low COP_H. By analogy with the double-effect cycles, they can be called "double-lift" cycles. Cycles H7 and H5 suffer from relatively high pressures if used with NH_3/H_2O. Nevertheless, H7 has been sold as a refrigeration cycle. In such an application, the condensation pressure is lower than in heat pump applications. All three cycles, H5 to H7, are well suited for refrigeration if very low temperatures are required.

Heat Transformer Cycles

Heat transformers are proposed for heat recovery in industrial processes. The reference system is the direct fired process, Figure 8.23a. If heat losses of the process are ignored, the rejected heat equals the input heat, $Q_{in,ref} = Q_{use}$. This output, Q_{use}, is supplied to the heat transformer

(Figure 8.23b), and the fraction COP_T of this heat is recovered. Additional heating, $Q_{in} = Q_{use}$ $(1 - COP_T)$, is required. For the energy-saving factor, the following expression is found:

$$e = COP_T \qquad (8.25)$$

Results are shown in Table 8.7 and in Figure 8.22.

The double-effect cycles T2 to T4, Figure 8.22, yield more energy savings with a smaller temperature lift than the single-effect cycle T1. If a high temperature lift is required, cycles T5 to T7 are suitable. Their energy savings are smaller than for low temperature lift cycles.

Heat Pump Transformer Cycles

One cycle of this kind has already been described above. In Table 8.8, two other cycles with different temperature lifts are compared with the best heat pumps and transformers which yield the same lift. The reference for e is direct firing of an industrial process (Figure 8.23a). Consequently, for the heat transformer cycles, e is equal to the COP_T. For the heat pump cycles (see Figure 8.23c), the energy saving amounted to $e = 1 - 1/COP_H$. For the heat pump transformer cycles (Figure 8.22d), the following is found with the definition of their COP_{HPT} $= Q_{use}/Q_{in}$:

$$e = 1 - 1/COP_{HPT} \qquad (8.26)$$

The energy-saving factor is displayed in Figure 8.24. Cycles T1, H1, and HPT1 all provide the same temperature lift. However, the heat pump transformer yields a considerably higher energy saving than the two simple cycles. The same holds for cycles T4, H3, and HPT2, where the temperature lift is smaller, but the COP_{HPT} and energy savings are appreciably higher than in the first three cycles.

8.4.4 Concluding Remarks

With the method based on the superposition principle, the COP can be estimated for complicated advanced absorption cycles within an accuracy of about 10%. The advanced cycles are considered to be composed of basic single-stage heat pumps and/or transformer cycles. Thus, the COP for an advanced cycle can be written as a function of the efficiencies of the basic cycles. The major irreversibilities of the advanced cycle are taken into account by using realistic efficiencies of the single-stage cycles. These efficiencies are the only data needed for the calculation. By comparing different advanced cycles and different working fluids, guidelines can be found for choosing an appropriate cycle for each application. The precision of the presented calculation method is sufficient to evaluate and compare different cycle designs, different working fluid pairs, and even to make estimations about the investment cost.

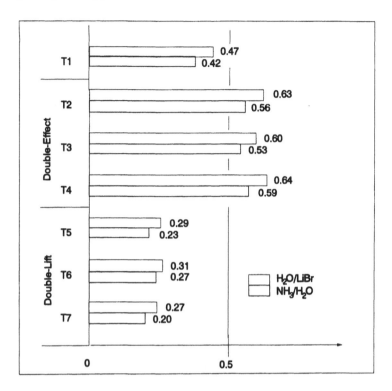

Figure 8.22 Coefficient of performance of several advanced heat transformer cycles.

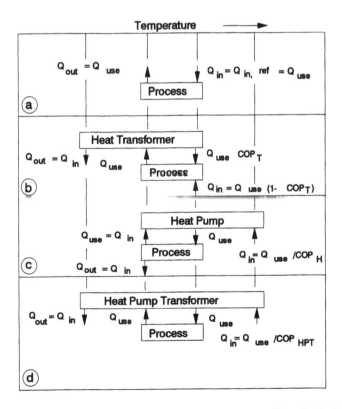

Figure 8.23 Different heat supply systems for a separation process. (a) direct fired, (b) heat transformer, (c) heat pump, (d) heat pump transformer.

Table 8.7   Double-Effect and Double-Lift Cycles Compared to a Single-Stage Heat Transformer and Direct Heating with Regard to COP_T

HEAT TRANSFORMER	COP_T	COP_T H_2O/LiBr
		COP_T NH_3/H_2O
T1	η_T	0.47
		0.42
T2	$\dfrac{\eta_T}{1-\eta_T+\eta_T^2}$	0.63
		0.56
T3	$\dfrac{\eta_T+\eta\eta_T}{1+\eta\eta_T}$	0.60
		0.53
T4	$\dfrac{2\eta_T}{1+\eta_T}$	0.64
		0.59
T5	$\dfrac{\eta_T^2}{1-\eta_T+\eta_T^2}$	0.29
		0.23
T6	$\dfrac{\eta_T}{2-\eta_T}$	0.31
		0.27
T7	$\dfrac{\eta\eta_T}{1+\eta\cdot\eta_T}$	0.27
		0.20

Table 8.8   Heat Pump Transformers Compared to Heat Pumps and Heat Transformers with Regard to COP and Relative Energy Saving, e

HEAT PUMP TRANSFORMER	COP	COP H_2O/LiBr / COP NH_3/H_2O	e(%) H_2O/LiBr / e(%) NH_3/H_2O
Medium Lift			
T1	η_T	0.47	47
		0.42	42
H1	$1 + \eta$	1.72	42
		1.55	35
HPT1	$\eta + \dfrac{1}{1 - \eta_T}$	2.61	62
		2.27	56
Small Lift			
T4	$\dfrac{2\,\eta_T}{1 + \eta_T}$	0.64	64
		0.59	59
H3	$1 + 2\eta$	2.44	59
		2.10	52
HPT2	$\dfrac{1 + \eta + \eta_T}{1 - \eta_T}$	4.13	76
		3.40	71

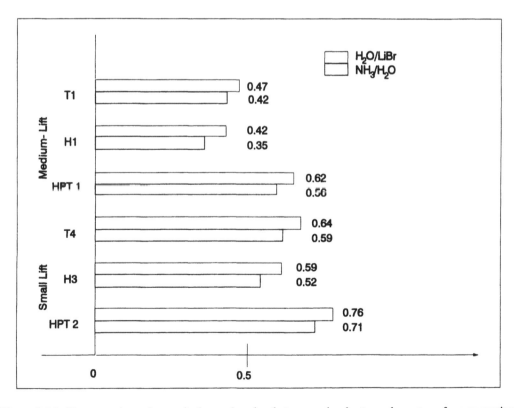

Figure 8.24 Energy savings of several advanced cycles that are used as heat supply systems for a separation process.

REFERENCES

1. Alefeld, G. (1983): Le Froid Au Service De L'Homme, Paris, France.

2. Alefeld, G. (1982a): Regeln für den Entwurf von mehrstufigen Absorptionswmepumpen. In: Brennstoff-Wärmeumwandlungssysteme, Skriptum zur Vorlesung am Institut für Technische Physik E 19, Physik-Department der Technischen Universität München.

3. Alefeld, G. (1982b): Kompressions-und Expansionsmaschinen in Verbindung mit Absorberkreisläufen. In: Brennstoff-Wärme-Kraft, Bd. 34, Nr. 3, S. 142.

4. Alefeld G. (1981): Der Wärme pumpen transformation Brennstoff-Wärme-Kraft, Vol 33, pg. 486-490.

5. Alefeld, G., Brandl, F., Opitz, U., Völkl, J., Ziegler, F., (1985): "Advanced cycles for the use of solar energy", Intern. Symp. on Thermal Application of Solar Energy, April 7-10, Hakone, Japan.

6. Alefeld, G. and Ziegler, F., (1985): "Advanced heat pump and air-conditioning cycles for the working pair, H_2O/LiBr: Industrial Applications", *ASHRAE Transact.*, Vol. 91, Part 2, p. 2072 - 2080.

7. Alefeld, G. and Ziegler, F. (1985): "Advanced heat pump and air-conditioning cycles for the working pair, H_2O/LiBr: Domestic and Commercial Applications", *ASHRAE Transact.*, Vol. 91, Part 2, p. 2062 - 2071.

8. Ziegler, F. and Alefeld, G. (1987): "Coefficient of performance of multistage absorption cycles", *Int. J. Refr.* Vol. 10, p. 285.

CHAPTER NINE

Two-Stage Configurations with Compression/Expansion Machines

In this chapter, configurations are introduced which are obtained by coupling absorption cycles with compression/expansion machines. The coupling of two basic cycles is explained first. Then combinations of two-stage absorption configurations and one basic compression cycle will be described. A more detailed classification will follow in Chapter Ten.

9.1 CLASSIFICATION OF TWO-STAGE ABSORPTION-COMPRESSION/ EXPANSION MACHINES

Figure 9.1 represents all combinations of a basic absorption cycle with a basic compression/ expansion cycle. Configurations of Figure 9.1a to m are obtained by coupling one exchange unit of the absorption cycle A with one exchange unit of the compression/expansion cycle K. With this coupling several different configurations are obtained with the following characteristics:

α   12 classes, Figure 9.1a to m:

    5 exchange units

    1 compression/expansion machine

    2 working fluid streams

    3 pressure levels

    5 temperature levels (for some classes, these can be reduced to two by applying Rule 3b)

By merging two exchange units of the basic absorption and the basic compression/ expansion cycle, the configuration of Figure 9.1n is obtained with the following characteristics:

β   1 class, Figure 9.1n:

    4 exchange units

    1 compression/expansion machine

    2 working fluid streams

    2 pressure levels

    4 temperature levels (which can be reduced to two)

By applying Rule 3a to the classes of the configurations of Figure 9.1a to m, two of the original classes are transformed into one new class, shown in Figure 9.1o to t, with the following characteristics:

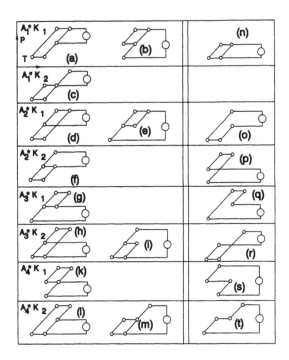

Figure 9.1 Two-stage compression/expansion-absorption machines.

γ   6 classes, Figure 9.1o-t

    4 exchange units

    1 compression/expansion machine

    1 working fluid stream

    3 pressure levels

    4 temperature levels (which can be reduced to three and for configurations o, p, and s to two by applying Rule 3b).

In Figure 9.1, the two basic cycles are heat and mass coupled. However, all of these configurations can be constructed as cascades as well.

Figures 6.3 and 6.4 (Chapter Six) show two examples corresponding to the configurations of Figure 9.1c. Most of the modes discussed in the following chapters can also be operated as cascades.

9.2 SPECIAL MODES OF OPERATION

The operating modes of the configurations from Figure 9.1n,f,i,k, and b are discussed in detail. Figure 9.2a to e show these five arrangements of absorption/compression machines. It is assumed that the units are operated with only three essential temperature levels. Exchange units located on the same temperature level may stay in internal heat exchange. This is not always required for the operating modes discussed in the following: Figure 9.3 shows the configuration of Figure 9.2a in more detail. The connection between exchange units B and D may be either an expansion device or a pump or both, as shown. There are three complete (in this case basic) cycles: WAC and WBD represent vapor-compression heat pumps if the working fluid circulates counterclockwise. The cycle ABDC represents a heat transformer. By

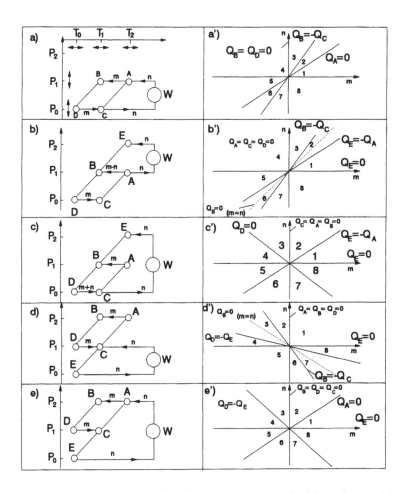

Figure 9.2 Examples of modes of operation of two-stage compression/absorption machines with three temperature levels.

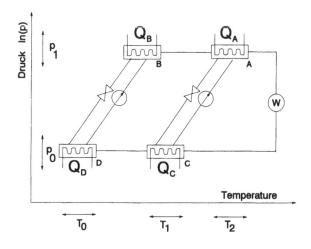

Figure 9.3 Coupling of a basic absorption cycle with a basic compression/expansion cycle.

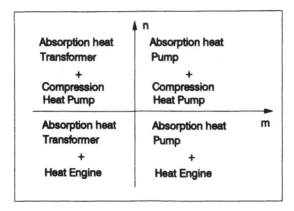

Figure 9.4   Operating modes for the configurations of Figure 9.1.

Q_i	Fig. 9.2	Fig. 9.2	Fig. 9.2	Fig. 9.2	Fig. 9.2
	a	b	c	d	e
Q_A	m-n-W	m	m	m	m-n-W
Q_B	-m	n-m	-m	-m	-m
Q_C	n-m	-m	-m	-m-n-W	-m
Q_D	m	m	m+n	m	m
Q_E	-	-n-W	-n-W	n	n
W	$n/(\eta-1)$	$n/(\eta-1)$	$n/(\eta-1)$	$n/(\eta-1)$	$n/(\eta-\eta)$
Q_2	$m-n\eta/(\eta-1)$	$m-n\eta/(\eta-1)$	$m-n\eta/(\eta-1)$	m	$m-n\eta/(\eta-1)$
Q_1	n-2m	n-2m	-2m	$-2m-n\eta/(\eta-1)$	-2m
Q_0	m	m	m+n	m+n	m+n

Where $n>0: \ \eta = \eta_{\text{heat pump}}$
       $n<0: \ \eta = 1/\eta_{\text{heat engine}}$

Figure 9.5   Amounts of work and heat exchanged.

combining any two of the three basic cycles, all components of Figure 9.3 are represented. Here the cycles WAC and ABDC are chosen to be independent, and the circulating working fluid streams are proportional to n and m. The configurations of Figure 9.2b to e can be treated in a similar manner.

The four sign combinations of the fluid streams m and n represent the four combinations of basic cycles shown in Figure 9.4. The signs and the approximate values of the heat exchanged in the respective exchange units A to F and of the heat Q_2, Q_1, Q_0 at the corresponding temperature levels T_2, T_1, and T_0 are displayed in Figure 9.5. The configurations of Figure 9.2b to e with five exchange units seem initially to be more complicated than the one in Figure 9.2a with four exchange units. However, as will be shown later, for several operating modes one of the exchange units in Figure 9.2b to e is small or vanishes entirely, resulting in a first cost that is not substantially increased compared to that of the configuration in Figure 9.2a. Each one of the configurations has specific characteristics. These are displayed next to each configuration in Figure 9.2 as a diagram for the operating modes. The boundaries of operating modes are defined by the conditions $Q_1 = 0$, $Q_2 = 0$, $Q_3 = 0$, and $W = 0$, which is equivalent to n = 0. The sign combinations of the heat and work exchanged are found in Figure 9.6. A comparison with Figure 1.9   (Chapter One) reveals that any one of the configurations in Figure 9.2a to e can be operated in all reversible modes which are possible according to the First and Second Laws.

1	2	3	4	5	6	7	8	
+	+	+	+	-	-	-	-	W
+	-	-	-	-	+	+	+	Q_2
-	-	+	+	+	+	-	-	Q_1
+	+	+	-	-	-	-	+	Q_0

Figure 9.6 Sign combinations of operating ranges that may be reversible.

In the following, the modes of operation and the ranges in the m, n plot will be discussed for the configuration of Figure 9.2a. Eight special cases are represented by the boundaries between the Operating Ranges 1 through 8 of Figure 9.2a':

n = 0   This case represents absorption operation only, either as heat pump (m > 0) or heat transformer (m < O). The compression/expansion machine is not in operation;

m = 0   This case represents $Q_0 = 0$ and $Q_B = 0$. It is a vapor-compression heat pump (n > 0) or expansion machine (n < 0) with components A,C,W and a solution circuit operating between temperatures T_1 and T_2;

m = n   This case represents $Q_2 = Q_A = 0$ and $Q_C = 0$. It is a vapor-compression heat pump (n > 0) or expansion machine (n < 0), operating with components B,D,W between temperatures T_0 and T_1. The solution circuit is optional.

n ≈ 2m   This case represents $Q_1 = 0$, $Q_B = -Q_C$. It is a vapor-compression heat pump (n > 0) with internal heat exchange between B and C [or an expansion machine (n < 0)] operating with a large temperature lift from T_0 to T_2 at low pressure ratio P_1/P_0. A single-stage vapor-compression heat pump would have to operate between the pressure levels P_0 and P_2.

The eight operating ranges exhibit the following characteristics:

Operating Range 1: The driving heat for the absorption heat pump is supplied to A, useful heat is generated at temperature T_1, and cooling is provided at T_0. For n << m, the capacity of the absorption heat pump dominates; for n → m the capacity of the vapor-compression heat pump dominates.

Operating Range 2: The compressor operates between the exchange units BD and AC simultaneously. Heat is rejected from A at T_2 and from B at T_1. Cooling is produced by D at T_0. The desorber heat required by C may be supplied by internal heat transfer from D or from an external source. The average temperature level of the useful heat increases from T_1 (n = m) to T_2 (n ≈ 2m), with decreasing COP. This cycle is well suited as a heat pump to be used for heating or as an air-conditioner employing a variable temperature level for heat rejection. The same holds for cooling. This is possible because of the flexibility of the heating temperature level (or waste heat temperature level) to respond to varying operating conditions. Figure 9.7 shows a detailed configuration in which the exchange units B and C are merged into a three path heat exchanger. If the compressor operates mainly on exchange units B and D (boundary between the operating ranges 1 and 2), the superheat can be used in A to desorb working fluid vapor. This improves the efficiency further. The vapor generated in D may bypass C to avoid further superheating. The vapor generated in C may even be precooled in D to enter the compressor with lower superheat. The determination of the optimum vapor temperature entering the compressor requires a more detailed consideration not discussed here.

Operating Range 3: The heat released at B cannot completely satisfy the heat requirement of C. Therefore, cooling is provided on two temperature levels: at T_0 (in D) and at T_1 (in C). Heat is rejected by A.

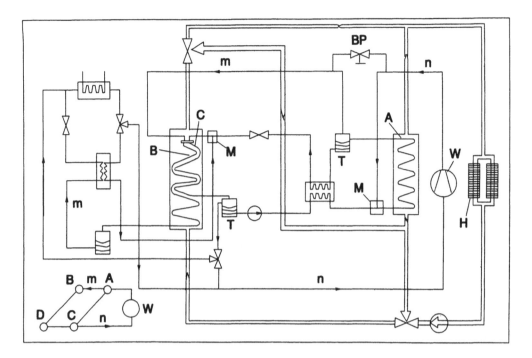

Figure 9.7 Vapor compression heat pump of Figure 9.2a in operating range 2. B, C, three pass heat exchanger; BP, bypass; H, heating system; M, device for mixing of vapor and liquid phases; T, separator of vapor and liquid phase.

Operating Range 4: The configuration operates as a combination of a heat transformer and a vapor-compression heat pump. Waste heat is rejected at T_0, heat is supplied to B and C at T_1 as a heat source for the heat transformer, and additional heat is supplied at C as a heat source for the vapor-compression cycle, while useful heat is provided at T_2.

Operating Range 5: This configuration is a heat transformer that simultaneously delivers work. Waste heat is rejected at T_0, heat is supplied at T_1 to B and C, and useful heat is provided at T_2. Work is produced in W.

Operating Range 6: This configuration produces work from heat supplied at T_1 to B and T_2 to A. Heat rejected at C is transferred to B. Waste heat is rejected at T_0 only.

Operating Range 7: The configuration produces work from heat supplied at T_2 and has internal heat exchange from C to B; waste heat is rejected at T_0, and heat is rejected from C as waste or useful heat.

Operating Range 8: This configuration is an absorption heat pump, ABCD. Part of the working fluid vapor generated in A is expanded in W and absorbed in C. Heat is supplied at T_2, rejected at T_1 (B and C), and cooling is provided at T_0. This configuration produces cooling and/or work dependent on the ratio n/m.

Operating ranges 1 to 8, including the eight special cases marking the boundaries of the operating ranges, exist for all other configurations of Figure 9.2b to e, too. However, the compression/expansion machines operate on higher or lower pressure levels or pressure ratios.

In the following, only a few operating modes will be explained, which represent special cases characterized by a vanishing heat exchange to an outside sink or source on a certain temperature level. These cases are represented in the diagrams by the boundaries between two segments of Figure 9.2a' to e'. These operating modes are obtained by internal heat exchange between units located at the same temperature level or by leaving parts of the cycles unused.

The following sections marked A are concerned with the production of useful heat, sections B are concerned with the production of cooling, and sections C are related to power generation (the lower half-plane of the diagrams):

$Q_1 = 0$ represents the boundary between operating ranges 2 and 3, and 6 and 7:

A. Heat is pumped from T_0 (for example, the temperature of the surroundings) to T_2. In configurations 9.2a,b,d, and e, the compressor operates on higher or lower pressure levels and pressure ratios than those which would be obtained with the pure working fluid only. The internal heat transfer at T_1 produces a large temperature lift.

B. Refrigeration is available at a relatively low temperature T_0 when T_2 represents the temperature of the surroundings. A large temperature lift is obtained here also. The pressures show the same characteristics as described under A.

C. Work is produced from heat supplied at T_2, and waste heat is rejected at T_0. However, the pressure ratio across the expansion machine is much smaller for configurations 9.2a,b, and d as compared to the pressure ratio of a pure working fluid. The absolute value of the high side pressure for the configuration 9.2a,d, and e is lower than P_2 (which is the pressure of the pure working fluid at T_2). The absolute value of the low side pressure is higher than P_0 (which is the pressure of the pure working fluid at T_0) for the configuration of Figure 9.2b. It is smaller than P_0 for the configuration of Figure 9.2d and e.

$Q_2 = 0$ represents the boundary between operating ranges 1 and 2, and 5 and 6:

A. Heat is pumped from T_0 to T_1. In the configurations of Figure 9.2b and d, the compressor faces the same pressure ratios as with the common heat pump using a pure working fluid, but the absolute pressure may be higher (configuration 9.2b) or lower (configuration 9.2d). For configuration 9.2c and e, the pressure ratio is larger than for a common, pure working fluid heat pump.

B. Cooling is produced at T_0; waste heat is rejected at T_1, and the COP is doubled compared to the operating mode where $Q_1 = 0$ (because of the low lift). The pressures show the same characteristics as mentioned under A.

C. Heat of temperature T_1 is converted into work. The waste heat is rejected at T_0. For the configurations of Figure 9.2b to e, the expansion machine can be operated in a different pressure and temperature range than by using the pure working fluid only.

$(Q_0 = 0)$ represents the boundary between 3 and 4, and 7 and 8:

A. Heat is pumped from T_1 to T_2. However, the compressor operates in a pressure range different from the one for a conventional unit with the same working fluid. This applies to all configurations except 9.2b.

B. Cooling is produced at T_1; waste heat is rejected at T_2, and the compressor operates again in a different pressure range.

C. Heat is supplied at T_2 and is converted into work. Waste heat is rejected at T_1. With regard to the pressure range, the same situation as in A is encountered again.

Example 9.1: Vapor-Compression Heat Pump with Two Solution Circuits

A vapor-compression heat pump is evaluated that employs two solution circuits and operates in operating range 2. A detailed sketch of the cycle is shown in Figure 9.8. There is no heat rejected at the intermediate temperature level. The entire heat released by B is used in C. Cooling is provided by D

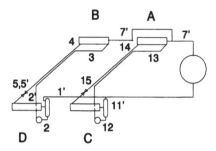

Figure 9.8 Vapor-compression heat pump with two solution circuits.

and heating by A. The vapor leaving D is mixed with the vapor from C and fed into the compressor. Since the vapor concentrations are different, the vapor leaving C has to be rectified to remove excess water. Otherwise water would be carried over from the high temperature solution circuit to the low temperature solution circuit. All fluid streams leaving the exchange units are saturated. The effectiveness of the solution heat exchangers is 1.0, and the compressor efficiency is 1.0. The temperature of the fluids leaving D is 0°C, and the concentration of the liquid phase is 0.6. The overall concentration of the fluid entering D is 0.7. The temperature of the liquid leaving A is 100°C, and the concentration change in A is 0.1. To ensure a sufficient temperature difference for the heat transfer from B to C, the lowest temperature of B is set to be equal to the highest temperature of C. The vapor leaving C has to be rectified to the same concentration as the vapor leaving D. The condition of the vapor leaving the rectifier is assumed to be identical to the vapor leaving D. The heat of rectification is rejected to the fluid stream that is cooled in D.

(a) What are the values of all properties at all state points?

(b) What are the amounts of heat exchanged and the work requirement by the compressor and the pumps?

(c) What is the COP of the system?

Solution: (a) Figure 9.9 shows a schematic enthalpy-concentration diagram. The numbers refer to the same state points as listed in Figure 9.8. The properties are summarized in Table 9.1. The low temperature solution circuit is termed the first stage, and the high temperature solution circuit is termed the second stage. The state points of the second stage follow the same numbering system as for the first stage, with ten added to the value. The points are found as described in previous examples, especially Example 3.4 (Chapter Three). Care must be taken to determine the fluid conditions at the solution heat exchanger inlets and outlets. The fluid in one path cannot be heated further or cooled lower than the temperature of the stream entering the other path. The vapor leaving the desorbers is assumed to be in equilibrium with the leaving liquid streams.

(b) For the first stage, the amount of work provided to the solution pump is

$$w_p = (f-1)(P_3 - P_2) v_s$$

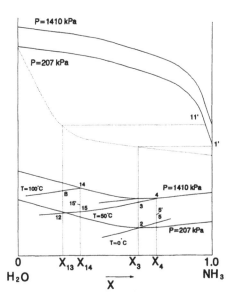

Figure 9.9 Schematic enthalpy-concentration diagram for configuration of Figure 9.8.

**Table 9.1 Thermodynamic Properties
at the State Points**

	T(°C)	P(kPa)	x	h(kJ/kg)
1′	0	207	0.999	1282
2	0	207	0.60	-238
2′	0	207	0.60	-237
3	50	1410	0.60	7
4	50	1410	0.70	45
5	0	207	0.70	-185
5′	14	1410	0.70	-138
6′	0	207	0.999	1282
7′	144	1410	0.999	1577
11′	50	207	0.96	1442
11″	0	207	0.999	1282
12	50	207	0.27	30
12′	50	207	0.27	31
13	100	1410	0.27	259
14	100	1410	0.37	226
15	50	207	0.37	-9
15′	57	1410	0.37	29

with

$$f = (x_{1'} - x_3)/(x_4 - x_3)$$

$$f = (0.999 - 0.6)/(0.7 - 0.6) = 4.0$$

and

$$w_p = (4.0 - 1)(1410 \text{ kPa} - 207 \text{ kPa})\, 1/840 \text{ m}^3\text{kg} = 4.3 \text{ kJ/kg}$$

With this, the enthalpy $h_{2'}$ amounts to $h_{2'} = h_2 + w_p/(f-1) = -237$ kJ/kg. Then the heat capacity of both streams, q_r and q_p, in the solution heat exchanger is calculated.

$$q_{rs} = (f-1)(h_3 - h_{2'}) = 3.0 \left(7 \text{ kJ/kg} - (-347 \text{ kJ/kg})\right) = 732 \text{ kJ/kg}$$

and

$$q_{ls} = f(h_4 - h_5) = 4.0 \left(45 \text{ kJ/kg} - (-185 \text{ kJ/kg})\right) = 920 \text{ kJ/kg}$$

The enthalpy $h_{5'}$ amounts to $h_{5'} = h_4 - q_{rs}/f = -138$ kJ/kg. The temperature of the stream leaving the solution heat exchanger at point 5′ is 14°C. The heat of desorption in D amounts to:

$$q_D = h_{1'} - h_2 + f(h_2 - h_{5'})$$

$$q_D = 1282 \text{ kJ/kg} - (-237 \text{ kJ/kg}) + 4.0 \left((-237 \text{ kJ/kg}) - (-138 \text{ kJ/kg})\right)$$

$$q_D = 1123 \text{ kJ/kg}$$

The compressor work is calculated based on the assumption that the vapor is pure ammonia by using the pressure-enthalpy chart of pure ammonia. Since the condition of point 6 is not yet known, it is assumed that vapor of point 1′ enters the compressor. The compressor discharge, point 7, is determined by the inlet conditions and the isentropic compression process that ends at a pressure level of 1410 kPa.

$$w_{comp} = 825 \text{ kJ/kg} - 530 \text{ kJ/kg} = 295 \text{ kJ/kg}$$

The enthalpy values in the above equation are read from the pressure-enthalpy chart for pure ammonia and shifted by a constant value compared to the values of the enthalpy-concentration diagram. The enthalpy of the vapor at the compressor outlet is $h_{7'} = h_{1'} + 295$ kJ/kg $= 1577$ kJ/kg. The heat of absorption in B amounts to:

$$q_B = h_1 - h_3 + f(h_3 - h_4)$$

$$q_B = 1577 \text{ kJ/kg} - 7 \text{ kJ/kg} + 4.0 \left(7 \text{ kJ/kg} - 45 \text{ kJ/kg}\right) = 1418 \text{ kJ/kg}$$

The same calculation procedure is now repeated for the second stage. There are two complications. The vapor leaving C has to be rectified to the same concentration as the vapor leaving D. That implies that the vapor leaving the rectifier has the same temperature, concentration, and pressure as the vapor leaving D. Based on 1 kg of vapor produced in C, the following values are obtained. The subscript ss refers to parameters of the second stage:

$$f_{ss} = (0.999 - 0.27)/(0.37 - 0.27) = 7.3$$

$$w_{pss} = 6.3 \,(1410 \text{ kPa} - 207 \text{ kPa})\, 1/840 \text{ m}^3\text{kg} = 9.0 \text{ kJ/kg}$$

$$h_{12'} = h_{12} + w_{pss}/(f_{ss} - 1) = 31 \text{ kJ/kg}$$

$$q_{rsss} = 6.3 \ (259 \text{ kJ/kg} - 31\text{kJ/kg}) = 1436 \text{ kJ/kg}$$

$$q_{lsss} = 7.3 \ (226 \text{ kJ/kg} - (-9\text{kJ/kg})) = 1716 \text{ kJ/kg}$$

$$h_{15'} = 226 \text{ kJ/kg} - (1436 \text{ kJ/kg})/7.3 = 29 \text{ kJ/kg}, 57°C$$

The amount of reflux in the rectifier is

$$v = (x_{11''} - x_{11'})/(x_{11'} - x_{12})$$

$$v = (0.999 - 0.96)/(0.96 - .27) = 0.057$$

$$q_r = (h_{11'} - h_{11''}) + v(h_{11'} - h_{12})$$

$$q_r = (1442\text{kJ/kg} - 1282\text{kJ/kg}) + 0.057 \ (1442\text{kJ/kg} - 30\text{kJ/kg}) = 240\text{kJ/kg}$$

The compressor work is the same as for the first stage. The desorber heat for C amounts to:

$$q_C = 1282 \text{ kJ/kg} - 31 \text{ kJ/kg} + 7.3 \ (31 \text{ kJ/kg} - 29 \text{ kJ/kg}) + 240 \text{ kJ/kg} = 1506 \text{ kJ/kg}$$

and the absorber heat for A amounts to:

$$q_A = 1577 \text{ kJ/kg} - 259 \text{ kJ/kg} + 7.3 \ (259 \text{ kJ/kg} - 226 \text{ kJ/kg}) = 1559 \text{ kJ/kg}$$

The amounts of heat q_B and q_C have to be equal. To achieve this, the mass flow rate in the second stage will be adjusted by a factor of $q_B/q_C = (1418$ kJ/kg)/(1506 kJ/kg) = 0.942. The amounts of heat and work of the second stage have to be weighted with this factor to obtain a match of the internal heat exchange between B and C. For the entire cycle, the following values are obtained:

Pump work:	$w_p + w_{pss}$ 0.942 = 12.8 kJ/kg
Compressor work:	(1 + 0.942) w_{comp} = 573 kJ/kg
Absorber heat A:	q_A 0.942 – 1469 kJ/kg
Desorber heat D:	q_D-q_r 0.942 = 897 kJ/kg

Internal heat exchange: $q_B = q_C$ 0.942 = 1418 kJ/kg.

The energy balance yields 573 kJ/kg + 13 kJ/kg + 897 kJ/kg-1469 kJ/kg = 15 kJ/kg

(c) The COP amounts to:

$$COP = 897/573 = 1.57$$

Remarks: 1. The cycle discussed in this example provides a very high temperature lift and is a competitor to the conventional cascade of two refrigeration cycles. However, here only one compressor is needed.

2. To increase the efficiency of the system it is recommended to replace rectifier with bleedline from point 3 to 15 in Figure 9.8.

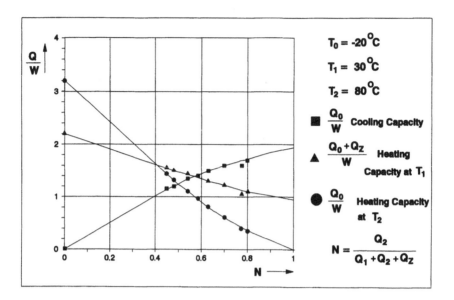

Figure 9.10   Measured coefficients of performance as a function of the ratio N of the useful heat.

3. In this calculation, it is assumed that the vapor for B bypasses A. This is not necessary. If the vapor were passing through A, it could be stripped of part of its water content, reducing the loss due to rectification.

4. The pressure ratio of the system investigated here is smaller than for a conventional system of the same temperature lift. This is a consequence of the staging.

9.3   ADJUSTMENT OF COEFFICIENTS OF PERFORMANCE AND LARGE TEMPERATURE LIFTS

The configurations of Figure 9.2 exhibit the important characteristic that the operating modes can be adjusted to varying external conditions by merely changing the fluid streams n and m.

As an example, operating range 2 is considered once more. All configurations of Figure 9.2 can be operated in range 2. Useful heat can be extracted at T_2 and T_1. The COP increases as the heat production shifts from T_2 to T_1. In Figure 9.10, experimental data are shown for a machine, according to Figure 9.7, operated in range 2. Q_0, Q_1, and Q_2 are the amounts of heat transferred externally at T_0, T_1, and T_2. Q_z is the heat rejected by the compressor. N gives the ratio of the heat extracted at T_2 in relation to the total heat extracted from the cycle. In Figure 9.10, the amounts of heat are plotted for one unit of work input. N equal to zero means extraction of heat at T_1 only. N equal to one would correspond to heat extraction at T_2 only. The solid lines are theoretical curves. With Q_2 increasing from zero to 1.9, the COP for cooling decreases from 2.2 to about 0.9. Accordingly, the heating capacity per unit work at T_1 decreases from 3.2 to 0. When heat is only rejected at T_2, the heating capacity for a given flow rate through the compressor is reduced (m ≈ n/2, Figure 9.2a), compared to the case in which all heat is being rejected at T_1 (m = n, Figure 9.2a). To maintain a constant heating capacity or to obtain an increase in heating capacity (while shifting the heating temperature from T_1 to T_2), the working fluid flow rate has to be increased. Thus, a variable speed compressor is required. The ratio of heat delivered at T_1 to the heat rejected at T_2 can be determined by controlling the temperature of the supply water (for example in a hydronic heating application).

The three path heat exchanger B/C from Figure 9.7 serves two purposes: internal heat exchange between B and C and heating of the external water stream, as required.

For refrigeration units and air-conditioners, the flexibility of the temperature level of the waste heat is of advantage. In case of a decreasing outdoor temperature, the unit achieves higher cooling COPs.

The configurations are capable of producing cooling capacity at two temperature levels simultaneously by employing one compressor only, as described in operating range 3. This feature is quite convenient for the design of refrigerators, freezers, and air-conditioners.

9.4 ADJUSTMENT OF PRESSURE LEVELS

Several examples are presented to emphasize the fact that the configurations described in Chapter Nine are capable of operating in pressure ranges which may be remarkably different from those encountered when only the pure working fluid without an absorbent is used.

Example 1: The configuration of Figure 9.2b in the operating range $Q_2 = 0$, $W > 0$ represents a vapor-compression heat pump lifting heat from T_0 to T_1 (temperature levels correspond to the pressure levels with the same subscript). The compressor is operating between P_1 and P_2 instead of between P_0 and P_1, which would be the case for a pure working fluid. The heat available from E is used (similar to mechanical vapor recompression) to generate vapor from A. Using water as the refrigerant and a 60%-weight LiBr solution as an absorber, the following operating conditions are obtained: $P_0 = 0.2$ bar, $P_1 = 1$ bar, $P_2 = 6.5$ bar, $T_0 = 60°C$, $T_1 = 100°C$, $T_2 = 150°C$. A conventional system operating with water and achieving the same temperature lift from 60°C to 100°C would have to handle large vapor volumes since the pressures would be 0.2 bar and 1 bar, respectively. In this example, the compressor operates between 1 bar and 6.5 bar.

Example 2: In the reversed case, $W < 0$, the described unit allows the use of the temperature difference between 100°C and 60°C for power generation with a small turbine by expanding steam from 6.5 bar to 1 bar. The absorber heat rejected by A will be used to generate steam in E.

The gliding temperatures in exchange units B and C are advantageous when the heat is rejected to a heat transfer fluid which does not undergo a phase change, but rather changes its temperature. The COP for the generation of work potentially increases beyond the one of a Rankine Cycle operating with a single component working fluid.

In both cases, T_0 can be chosen to be lower than 60°C. In both examples, the heat and mass transfer in exchange unit B is small, since for $Q_2 = 0$ stream m is approximately equal to stream n. When m = n, exchange unit B may be omitted entirely. In this case, Q_2 may only vanish completely for special conditions depending on heat capacities, heat of solutions, solution circulation rate, and compressor efficiency. The boundary between operating ranges 1 and 2 (Figure 9.2b′) may be located to the left or right of the dashed line for m = n or coincide in some intervals. If no heat is supplied or rejected to or from the unit at T_2, then exchange unit B [which operates as a small condenser (m > n, $Q_2 = 0$) or evaporator (m < n, $Q_2 = 0$) depending on the location of the two lines for $Q_2 = 0$ and $Q_B = 0$ with respect to each other] serves the purpose of matching the heat balance between A and E. The fluid stream m-n through B may be used to control the operating conditions of the combination WEA. The exchange units A, E, and B may be shifted along the vapor pressure lines to higher or lower pressures (see Figure 9.16a and c), so that for a given temperature lift from T_0 to T_1, the corresponding pressures P_2 and P_1 obtain technically more desirable values. To increase the efficiency, it is recommended to expand the liquid working fluid on its way from E to D first to P_1 and to recirculate the generated vapor into the compressor W.

For the boundary between operating ranges 5 and 6, an analogous case is found regarding the heat balance and component B. Since in this operating range liquid working fluid which has to be preheated is pumped from D to E, B may reject its condenser heat to this fluid stream, in the same way as feed water preheating is accomplished in conventional power plants.

The configuration of Figure 9.2d (and especially 9.16b) is especially useful when the thermal stability of the working fluid is affected by the superheat resulting from the compression process, or when the pressure level P_2 is inconveniently high. In the case of heat transformation-heat pump operation (when $Q_D = -Q_E$, that is $Q_0 = 0$, second quadrant in Figure 9.2d), heat is lifted from T_1 to T_2, although the compressor is operating between P_0 and P_1. For the operating condition $Q_0 = 0$ in the configuration of Figure 9.2d, the transfer of heat in the exchange unit C is small. The boundary between operating ranges 3 and 4, and 7 and 8 is located close to the dashed line $Q_C = 0$. Consequently, C may be omitted entirely, although the heat balance between D and E is not always fulfilled. Usually, exchange unit C will be kept to use the superheat of the gas leaving the compressor to generate further vapor, increasing the efficiency. In addition, C may be used to balance the solution heat exchange of the solution circulating between A and E. The exchange units D, E, and C may be shifted arbitrarily along vapor pressure lines to obtain a desirable pressure range (see Figure 9.16b and d). T_0 may be less than the temperature of the surroundings.

The configuration of Figure 9.2e represents a vapor-compression heat pump (under the operating conditions $Q_2 = 0$, m = n, and W>0) of high COP. Component A may be omitted (Figure 9.1s) unless the superheat is used in A for desorption purposes. If, instead of a compressor, an expansion machine is used (W < 0), A may serve as a superheater. The exchange unit connected to the discharge side of the compressor may act as an oil separator in addition to desuperheating. This is especially of interest when the absorbent is an oil suitable for compressor lubrication.

In the configuration of Figure 9.2c, exchange unit D may be omitted for $Q_0 = 0$, that is when m = −n.

9.5   ADJUSTMENT OF TEMPERATURE LIFT

For configurations of Figure 9.2a to f the temperature intervals T_2-T_1 and T_1-T_0 have been assumed to be equidistant on the inverse temperature scale. Figures 9.11a and b show how the temperature intervals can be changed to meet changing requirements, by including two additional exchange units and a solution circuit without increasing the number of essential temperature levels beyond three. Similar extensions are possible for any other absorption cycle of Figure 9.2a to e. The extended configurations which incorporate a two-stage absorption cycle can operate in the same modes as Figures 9.6 and 9.2a' to e'. The efficiencies will change. This is plausible because of the modified temperature intervals. In Figures 9.11a and b, each pressure level can be connected to any other by a compression/expansion machine.

9.6 SIMULTANEOUS COMPRESSOR AND ABSORPTION OPERATION

All configurations of Figure 9.2a to e contain two independent basic cycles, enabling the simultaneous supply of driving heat and work (with exception of special operating modes.) For example, these units may be operated initially by supplying heat only for absorption heat pump or absorption heat transformer operation. To increase the capacity, the compressor may be added. If the heat supply is exhausted, the compressor drives the entire unit. These configurations are especially of advantage when the compressor is driven by an internal combustion engine. The waste heat of the engine can be used to fire the absorption heat pump part (m > 0) or the absorption heat transformer part (m < 0). This applies also to the configuration of Figure 9.2b (and 9.16a), if water is the working fluid and the aqueous LiBr solution is the absorber (see Figure 10.11, Chapter Ten).

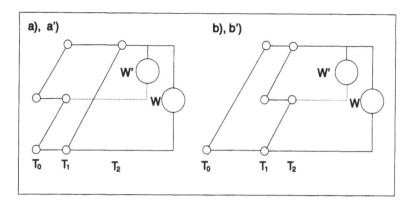

Figure 9.11　Vapor compression system with three solution circuits.

9.7 SPECIAL OPERATING RANGES FOR FOUR AND MORE TEMPERATURE LEVELS

Figures 9.16a,a' to k,k' show compression/absorption configurations with the exchange units arranged such that four essential temperature levels are obtained. To classify the operating modes of these arrangements, several configurations are discussed first which allow realization of all possible operating ranges for four temperature levels. Figure 9.12 shows the sign combinations for the heat Q_3 through Q_0 for the temperature levels T_3 to T_0, which are consistent with a reversible operation, including the exchange of work ($W \neq 0$) (see Chapter One). In Figure 9.12, columns located side by side differ by one sign only. At the boundary, the transition from one sign vector to the next, the heat that switches the sign becomes zero, corresponding to a simpler operating mode. In Figure 9.12, the sign combinations for all Q_i of the operating modes 1 through 4 are identical to those of 16 through 19, and those of 5 through 8 are identical to those of 12 through 15. Therefore, these operating modes are also possible for $W = 0$, i.e., no exchange of work. The sign combinations of the Q_i in the operating modes 9 through 11 and 20 through 22 are reversible only with the exchange of work. The three configurations of Figure 9.13 a,b_1, and b_2 with six exchange units and one compression/ expansion machine are composed of three basic cycles. Three fluid streams can be chosen independently. These configurations can be operated in all 22 operating ranges of Figure 9.12, provided that internal heat exchange can be established as needed. Figure 9.14 shows the amount of heat exchanged in the exchange units of the configurations of Figure 9.13 as a function of the fluid streams n, m, and l. Figures 9.15a and b show the 22 operating ranges of the configurations of Figure 9.13a and b_1 (for $n_1 = n$, $n_2 = 0$) as a function of the fluid streams m and l. Figure 9.15a represents cycles using compressors. Figure 9.15b represents cycles that produce work. Special operating modes are given by the limits for $Q_i = 0$ and especially in the corners, that is $Q_i = 0$ and $Q_j = 0$. In the first case, one heat becomes zero, configurations are obtained for which two fluid streams can be chosen arbitrarily. In the second case, two amounts of heat become zero, all heat and mass transfers are determined by fixing the amount and direction of one fluid stream only.

All configurations of Figure 9.16a,a' to k,k' which are composed of five exchange units and one compression/expansion machine, are included as special cases in the configurations of Figure 9.13b_1 and b_2. However, it turns out that the discussion of the configurations of Figure 9.16a,a' to k,k' becomes simpler when these arrangements are considered to be generalizations of the configurations of Figure 9.2b to e. In the discussion of Figure 9.2, it was assumed for the sake of simplicity that the configurations are operated under such conditions that only three essential temperature levels are encountered. This does not have to be the case. The absorption cycle ABCD of configurations 9.2a to e may be shifted arbitrarily parallel to the temperature

	1	2	3	4	5	6	7	8	9	10	11	12	13	14	15	16	17	18	19	20	21	22
W	+	+	+	+	+	+	+	+	+	+	+	-	-	-	-	-	-	-	-	-	-	-
Q_3	+	+	+	+	-	-	-	-	-	-	-	-	-	-	-	+	+	+	+	+	+	+
Q_2	+	-	-	-	-	+	+	+	-	-	+	-	+	+	+	+	-	-	-	+	+	-
Q_1	-	-	+	+	+	+	-	-	-	+	+	+	+	-	-	-	-	+	+	+	-	-
Q_0	+	+	+	-	-	-	-	+	+	+	+	-	-	-	+	+	+	+	-	-	-	-

Figure 9.12   Operating ranges of absorption/compression machines with four temperature levels.

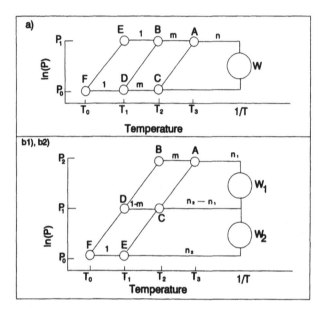

Figure 9.13   Three-stage absorption/compression machine.

axis or parallel to the vapor pressure lines. The only limitations are those imposed by the solution field. Further, the exchange units, E, of the compressor circuits may be shifted parallel to the vapor pressure lines. Figures 9.16 a,a' to d,d' show examples for how pressures, temperatures, and compositions of the exchange units in the configurations of Figure 9.2b to e can be modified by stretching the absorption cycle without affecting the topological relationship. This distortion has been selected so that four essential temperature levels are obtained. The possible operating ranges and operating modes of configurations of Figure 9.16a,a' to d,d' are different from those of Figure 9.2b to e. Yet they are included as subsets in Figure 9.12.

9.8   INTERNAL HEAT EXCHANGE

Since not all configurations of Figure 9.1 are considered in Figure 9.16, the various possibilities for internal heat exchange are listed in the tables of Appendix One, including internal heat exchange on one and on two temperature levels. The simultaneous internal heat exchange on two temperature levels can be balanced by installing a small additional exchange unit creating a third basic cycle and an additional independent fluid stream. The third stream is small compared to all other fluid streams. Of special interest are the configurations of Figure 9.1o,p, and s for which internal heat exchange is conducted, as shown in Figure 9.17. These three cases are included as special cases in Figure 9.2. Figure

Q_i	9.13 a	9.13 b_1	9.13 b_2
Q_A	m-n-W	m-n-W	m
Q_B	l-m	-m	-m
Q_C	n-m	1+n-m	l-m-n-W
Q_D	m-l	m-l	m-l
Q_E	-l	-l	n-l
Q_F	l	l	l
Q_3	m-nη/(η-1)	m-nη/(η-1)	m
Q_2	n+l-2m	n+l-2m	l-nη/(η-1)-2m
Q_1	m-2l	m-2l	m+n-2l
W_1	n/(η-1)	n/(η-1)	-
W_2	-	-	n/(η-1)

n>0: $\eta_{heat\ pump}$ n<0: $\eta = 1/\eta_{heat\ engine}$

Figure 9.14 Amount of heat exchanged in various exchange units as a function of the fluid streams.

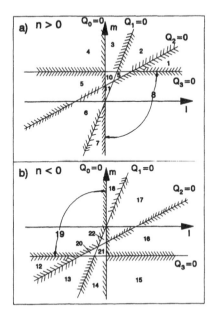

Figure 9.15 Operating ranges for configurations of Figure 9.12.

9.17a represents a vapor-compression heat pump, when operated counterclockwise, lifting heat from T_0 to T_1 while the compressor is operating in a much higher pressure range and the condenser heat is used for further desorption. In clockwise operation, this configuration represents a heat engine, with heat supply at T_1 and heat rejection at T_0, while the turbine

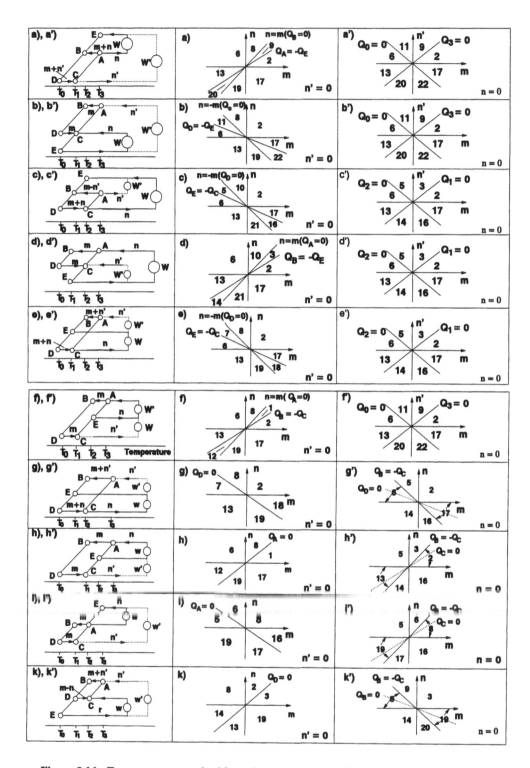

Figure 9.16 Two-stage compression/absorption machines with five (and four) exchange units.

is operating at a much higher pressure. The analog holds for the configurations of Figures 9.17b and c. For Figure 9.17c, the compressor or the expansion machine K operates at a considerably lower pressure than would be the case for a configuration without solution circuit A,C.

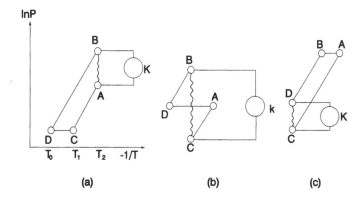

Figure 9.17 Compression/expansion absorption configurations with four exchange units and internal heat exchange.

The internal heat exchange indicated by the wavy line is not complete, resulting in the need for additional heat exchange with an outside reservoir. This heat exchange may be performed in a small additional exchange unit E and/or F, as shown, for example, in Figure 9.18.

9.9 REFRIGERATION, HEATING, AND POWER GENERATION

Generally, configurations are designed to serve one purpose only. The special feature of the two- and multistage configurations consists of the fact that two purposes can be fulfilled, simultaneously or alternatively. This flexibility, which is a consequence of the existence of two or more independent fluid streams, is especially advantageous when the heating and/or cooling demand is fluctuating while the power supplied remains constant. The three configurations of Figure 9.19a to c show, for example, how: (a) refrigeration capacity is provided at T_0 and work generated by supplying heat at T_2; (b) heating is provided at T_3 and work generated by supplying heat at T_2; and (c) heating and cooling is provided at T_3 and T_0, respectively, through input of work. The three applications of Figure 9.19 are examples of the general principle of combined heat, power and refrigeration.

9.10 USING SUPERHEAT

Usually the superheat generated by the compression of the working fluid is rejected within the condenser at a temperature level that is significantly lower than that of the superheated vapor, thus wasting availability. In multistage units, the amount of superheat may be reduced due to intercooling (see Section 9.2, above). Using the configurations of Figures 9.1e, it will be shown how the availability of the superheated vapor can be conserved partially by operating a small absorption cycle. In Figure 9.20, E is the evaporator and B is the condenser of a simple vapor compression heat pump. The exchange unit pair A/C is connected by a solution circuit. The vapor leaving the compressor rejects part of the superheat in A, desorbing working fluid. After the condensation in B, the liquid working fluid is partially expanded, and the generated vapor is separated from the liquid phase. Component D is a separator or receiver for liquid working fluid. The vapor is absorbed in C, and the absorber heat is rejected at T_1. Thus, the superheat is employed in a relatively small absorption cycle ABCD, the cooling capacity of which is used to subcool the liquid working fluid of the compressor cycle. This measure results in an increased capacity and increased COP for a given compressor, since the vapor generated in D does not have to pass through the compressor. The average composition of the solution circulating in A, C and, therefore, the temperatures of A and D are determined by the ratio of useful superheat to specific heat capacity available between B and D. At best, D may coincide with E, transferring the configuration of Figure 9.1e into the one of 9.1n. Also,

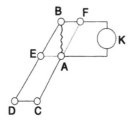

Figure 9.18 Complete internal heat exchange between A and B.

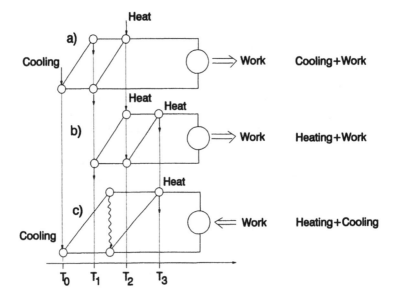

Figure 9.19 Simultaneous production of work, heat, and/or refrigeration.

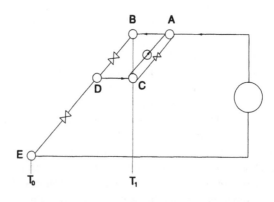

Figure 9.20 Configuration for using superheat.

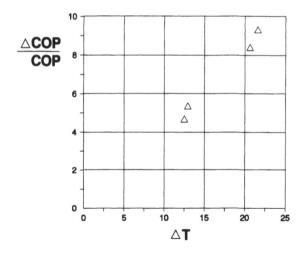

Figure 9.21 Measured improvement of the COP by utilizing the superheat according to Figure 9.20.

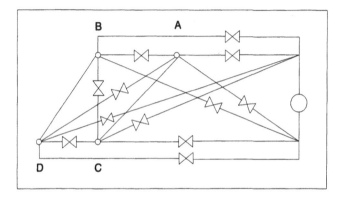

Figure 9.22 This configuration contains all configurations of Figure 9.1o to t.

in the configurations of Figure 9.1,a,b, and n, the availability of the superheat can be used. Figure 9.21 shows the improvement of the COP as it was measured for the configuration of Figure 9.20.

In the case of cycles generating work, it is often desirable to superheat the working fluid before entering the expansion process. The configuration of Figure 9.20 (with inversed fluid streams and pumps replaced by throttles and vice versa) may be used again. Heat is supplied at B and C. The vapor generated in B is superheated in A by absorbing a small portion of the vapor from B, while the vapor desorbed in C is used to preheat the liquid which is being pumped from E to D.

9.11 SWITCHING BETWEEN CONFIGURATIONS OF THE SAME CLASS

Absorber/compression configurations can be converted into each other by simply opening or closing valves of additional vapor lines in just the same way as shown previously for absorption cycles. The configurations of Figures 9.22 and 9.23 (symmetrical representation of Figure 9.22) include all configurations of Figures 9.1o to t and n. To switch between two classes, fewer valves are required, as shown in Figure 9.24c, for example. This configuration can switch between those of Figure 9.24a and b with internal heat exchange. In a similar way, the two sets of six classes each, S_1(a,b,c,k,l,m) and S_2(d,e,f,g,h,i), can be converted into each other just by opening and closing valves in vapor connections, as shown in Figure 9.25 for set S_2. Due to the switching, the pressures in the exchange units connected by vapor lines change,

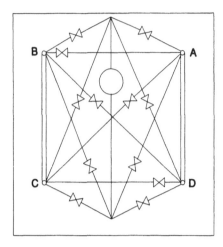

Figure 9.23   Switching between configurations of Figure 9.1n to t through valves in vapor lines.

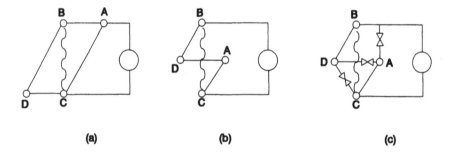

Figure 9.24   Switching between configurations with internal heat exchange.

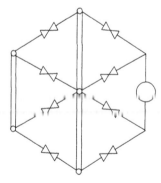

Figure 9.25   Switching between the configurations of Figure 9.1a to c,k to m.

and the temperature levels of those exchange units connected by solution circuits do change accordingly. The respective solution heat exchangers must be able to accommodate all desired operating modes.

9.12 SET PROPERTIES

In Figure 9.26, four elements are shown of which all configurations of Figure 9.1 can be composed. A line extended to the far right indicates a connection for a compression/expansion

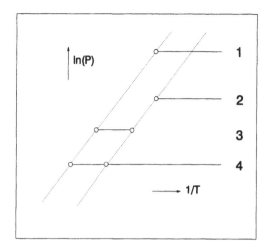

Figure 9.26 Elements for the construction of the configurations of Figure 9.1.

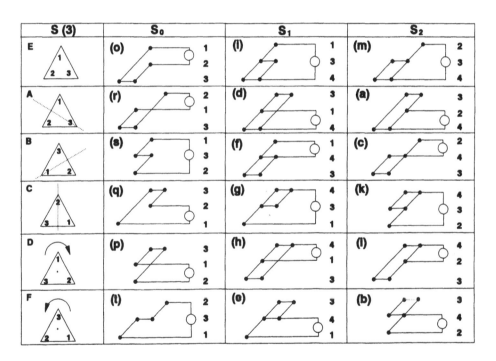

Figure 9.27 Application of the permutation operators A to F to the elements of Figure 9.24.

machine. As long as only one compression/expansion machine is used, only two such combinations of elements from Figure 9.26 are meaningful, for example combinations (1,2,3), (1,3,4) and (2,3,4). These three combinations and all their permutations are shown in Figure 9.27. As discussed in Chapter 7.11, the application of the six permutation operators E,A,B,C,D,F creates 18 configurations, according to Figure 9.1. The letters (a) through (t) in Figure 9.27 correspond to those in Figure 9.1. The configuration of Figure 9.1n is obtained by the combination (4,4). The conversions within sets and between the sets S_0, S_1, and S_2 of Figure 9.27 are again isomorph to the congruent operations of a tetrahedron of Figure 7.60 (Chapter Seven); but the plane 1,2,4 must be excluded when only two-stage arrangements are to be considered.

REFERENCES

See References in Chapter Three.

Multistage Absorption/Compression/ Expansion Machines

In this chapter, the discussion of combined absorption/compression machines is expanded to two- and multistage versions. The presentation begins with two-stage compression/expansion machines. Examples are heat engines, two-stage vapor-compression heat pumps, or combinations of a heat engine with a heat pump.

10.1 TWO-STAGE COMPRESSION/EXPANSION MACHINES

In Figure 10.1a to c, three configurations with two compression/expansion machines are shown. These are obtained according to Rule 1 by combining both exchange units of two basic cycles, as shown in Figure 5.3 (Chapter Five). The configuration of Figure 10.1d with only two exchange units and one independent fluid stream is obtained by applying Rule 3 to configurations a,b,or c.

The configuration of Figure 10.1d represents a two-stage power plant, if operated clockwise, or a two-stage vapor-compression heat pump, if operated counterclockwise. Usually the configuration of Figure 10.1b is used to improve the efficiency. In a power plant, a portion of the vapor is condensed after the first stage and preheats the liquid working fluid. In the case of a vapor-compression heat pump, the liquid working fluid is partially expanded, thus cooling the remaining liquid. The generated vapor is fed to the suction side of the high pressure stage. This reduces the flow rate through the low pressure stage, conserving work. In addition, the compressed vapor leaving the first stage is cooled by evaporation of a portion of working fluid on the intermediate pressure level reducing compressor work and superheat temperature.

If the two cycles of Figure 10.1b exhibit the opposite direction of fluid flow in each compression/expansion unit, a device is obtained with properties similar to an absorption heat pump or heat transformer. The work generated in one expansion machine may be used entirely to drive the compression machine. Such units, currently in an experimental stage, are called Rankine/Rankine Heat Pumps (Figure 10.2).

To limit the number of moving parts, expansion and compression machines may be combined to vapor a jet pump, as shown in Figure 10.3.

10.2 MULTISTAGE COMBINATIONS

The two-stage configurations of Figures 7.1, 9.1, and 10.1 can be expanded into three-stage configurations according to the following algorithms:

(A∗A)∗A (A∗A)∗K

(A∗K)∗A (A∗K)∗K

(K∗K)∗A (K∗K)∗K

A stands for a basic absorption cycle, and K stands for a basic vapor-compression cycle. The two-stage arrangements set in parentheses are expanded by one basic cycle, A or K, according to Rule 4. The application of Rule 4 is not commutative. The set of configurations obtained

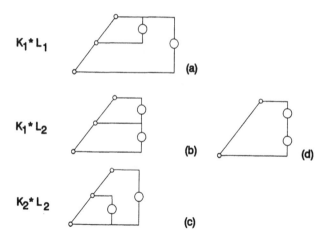

Figure 10.1 Two-stage compression/expansion machine configurations.

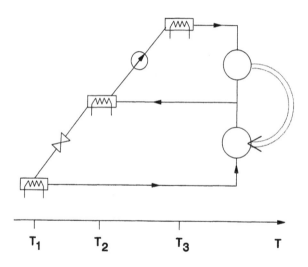

Figure 10.2 Rankine/rankine cycle heat pump.

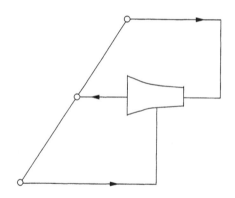

Figure 10.3 Schematic diagram of a vapor jet heat pump.

by applying the algorithm (K∗A)∗K is not the same as the set of configurations obtained by (K∗K)∗A. Depending on the sequence in which the two compression/expansion machines and the absorption cycle are combined, different configurations are obtained. It is recommended to begin the design of a multistage system with configurations composed of many exchange units (Figure 7.1a to t (Chapter Seven), Figure 9.1a to n (Chapter Nine), and Figure 10.1a to c). These will be expanded by a new basic cycle. Then the number of exchange units and independent fluid streams can be reduced according to Rule 3.

10.3 THREE-STAGE CONFIGURATIONS WITH ONE AND TWO COMPRESSION/EXPANSION MACHINES

These configurations are generated by applying Rule 4 according to the algorithms (A∗A)∗K and (A∗K)∗A. In this way, all possible configurations are obtained, although some configurations may be generated several times.

- A. When expanding configurations of Figure 7.1a to r by one basic compression cycle (which has one exchange unit in common with the original configuration, but which is not the common unit of the two original cycles), the following configurations are obtained when Rule 3a is applied zero to three times:

  8 exchange units, 3 streams, 1 compression/expansion machine

  7 exchange units, 2 streams, 1 compression/expansion machine

  6 exchange units, 1 stream, 1 compression/expansion machine

  The classes with six and seven exchange units can also be generated by expanding the configurations of Figure 7.1u to z by one compression cycle so that it has one exchange unit in common with the original configurations.

- B. When expanding the configurations of Figure 7.1a to r by a basic compression cycle which has one exchange unit in common with the original configuration so that this common unit is also the common one of the original configuration, the following arrangements are obtained when Rule 3a is applied:

  8 cxchange units, 3 streams, 1 compression/expansion machine

  7 exchange units, 2 streams, 1 compression/expansion machine

  Rule 3a cannot be applied more than once to these configurations.

- C. When expanding the configurations of Figure 7.1s and t by a basic compression cycle which has one exchange unit in common with the original cycles, the following configurations are obtained when Rule 3a is applied not more than once:

  7 exchange units, 3 streams, 1 compression/expansion machine

  6 exchange units, 2 streams, 1 compression/expansion machine

  Rule 3a cannot be applied more than once to these configurations.

- D. When expanding the configurations of Figure 7.1a to r by a basic compression cycle which has two exchange units in common with the original configuration, the following arrangements are obtained after applying Rule 3a:

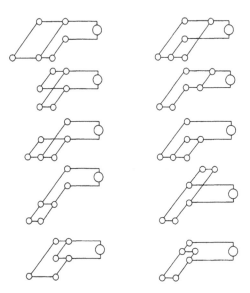

Figure 10.4   Absorption/vapor-compression system with six exchange units, one compression/expansion machine and two independent working fluid streams.

7 exchange units, 3 streams, 1 compression/expansion machine

6 exchange units, 2 streams, 1 compression/expansion machine

Rule 3a cannot be applied more than once to these configurations.

This subset of configurations with six exchange units, one compression/expansion machine, and two fluid streams can also be obtained by expanding the configurations of Figure 7.1u to z so that the compression cycle has two exchange units in common with the original cycles.

E.   When expanding the configurations of Figure 7.1s and t by one basic compression cycle which has two exchange units in common with the original arrangement, configurations with the following characteristics are obtained:

6 exchange units, 3 streams, 1 compression/expansion machine

Using the same scheme as listed under A to E, above, the configurations of Figure 9.1a to m or n, and o to t may be expanded by one absorption cycle. This reproduces, to some extent, configurations generated by the schemes A to E, but includes new configurations as well. Also, these configurations may be classified in the same groups A to E using the same relationships between exchange units, compression/expansion machines, and fluid streams. Examples of configurations with six exchange units, one compression/expansion machine, and two independent fluid streams are shown in Figures 10.4 to 10.9.

Summary

By applying the algorithms (A*A)*K and (A*K)*A, configurations with the following characteristics are obtained (arranged according to the number of independent fluid streams):

1 compression/expansion machine

3 working fluid streams, 8, 7 or 6 exchange units

2 working fluid streams, 7 or 6 exchange units

1 working fluid stream, 6 exchange units

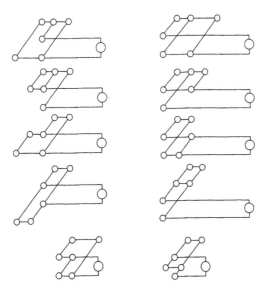

Figure 10.5 Absorption/vapor-compression system with six exchange units, one compression/expansion machine and two independent working fluid streams.

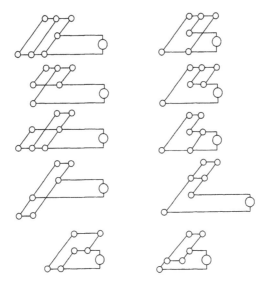

Figure 10.6 Absorption/vapor-compression system with six exchange units, one compression/expansion machine, and two independent working fluid streams.

For three-stage configurations with two compression/expansion machines, the following relationships between the number of independent working fluid streams and the number of exchange units hold:

2 compression/expansion machines

3 working fluid streams, 6 or 5 exchange units

2 working fluid streams, 5 or 4 exchange units

1 working fluid stream, 4 exchange units

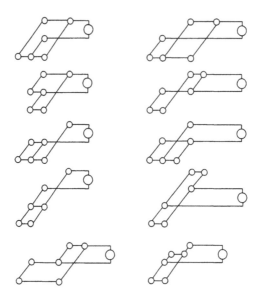

Figure 10.7 Absorption/vapor-compression system with six exchange units, one compression/expansion machine, and two independent working fluid streams.

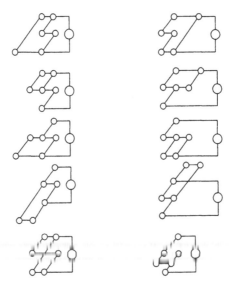

Figure 10.8 Absorption/vapor-compression system with six exchange units, one compression/expansion machine, and two independent working fluid streams.

10.4 SPECIAL MULTISTAGE CONFIGURATIONS

Of all working fluids discussed so far for absorption or vapor-compression heat pumps, only water offers the thermal stability required to take advantage of the availability of high temperature heat, as it is generated by combustion, by concentrating solar collectors, or by the flue gases of internal combustion engines.

On the other hand, the volumetric flow-rate of water and, consequently, of the compression/expansion machines becomes inconveniently large for low temperatures. Since large volumetric flow-rates do not impose a severe restriction on absorption machines, a configuration of Figure 10.10 might be convenient.

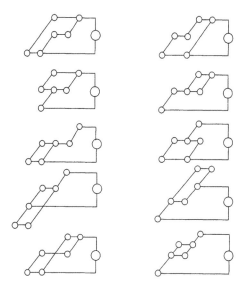

Figure 10.9 Absorption/vapor-compression system with six exchange units, one compression/expansion machine, and two independent working fluid streams.

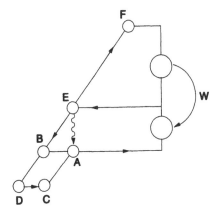

Figure 10.10 Schematic diagram of a configuration composed of one absorption cycle and a two-stage vapor-compression system.

A work-generating topping cycle with boiler F and condenser E is used to drive a compressor operating in a second cycle. The heat of condensation in E drives an absorption cycle, ABCD. B is a small unit, as described in Chapter 9.2. This configuration allows the use of the availability of high temperature heat for heat pumping or refrigeration. The corrosive or thermally unstable absorption solution reaches its highest temperature in A. Because the waste heat of the two compression/expansion machine provides the driving heat for the absorption cycle, the efficiency of the machines is not necessarily required to be optimum. The compression/expansion machines in Figure 10.10 may be replaced by a jet pump. If the work is generated by internal combustion, the configuration of Figure 10.11 can utilize the work output and waste heat of the engine.

Figure 10.12a shows how compressors can be integrated into the well-proven H_2O/LiBr-double effect machine. Figures 10.12b, c, and d display that the configurations of Figure 9.2a and b are used. For compressor K_2 the pressures range from 1 to 6 bar. If compressor K_1 or K_2 is driven by an internal combustion engine, waste heat may be used, depending on its

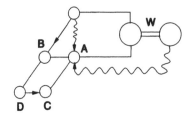

Figure 10.11 Schematic diagram of an absorption cycle driven by the waste heat of an internal combustion engine and the heat output of a compression system.

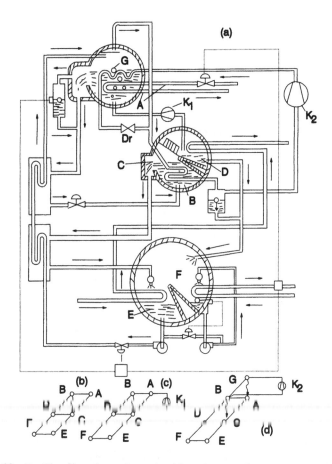

Figure 10.12 Double-effect absorption unit with an integrated vapor-compression system.

temperature level at A and/or C. Compared to common directly fired units ($T_A \approx 150°C$), the COP increases significantly as the following estimate suggests:

Assuming that 33% of the energy supplied is available as work (COP of the vapor-compression part is assumed to be approximately 3), that 27% is in the exhaust gas (used by A, COP for direct-fired double-effect units is approximately 0.9), that 25% is in the cooling water (used by C, COP for single-effect about 0.6), and that 15% heat losses are occurring, the overall COP for cooling amounts to 1.4. This value compares quite favorably to 0.9 for directly fired units.

The heat pump, which is used as a topping cycle according to Figures 10.10 to 10.12, may be operated in a cascade configuration using the same or different working fluids in the cycles.

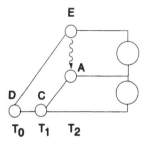

Figure 10.13 Vapor-compression system with internal heat exchange. The second compression/expansion machine is used to balance the internal heat exchange.

The increased flexibility that cascades provide has to be traded in for larger heat exchange areas.

Chapters 7.3 and 9.2 show how the internal heat exchange can be balanced by installing small exchange units. With these, a new independent cycle is constructed. The fluid stream of this cycle is adjusted to obtain the balance. Instead of an additional exchange unit, a small compression/expansion machine can be added, thus installing an additional compression cycle. By selection of the fluid flow rate, another condition, e.g., complete heat exchange at a certain temperature level, can be fulfilled. Figure 10.13 corresponds to Figure 9.2b. The operating state $Q_E = -Q_A$ can be accomplished by a low pressure compression/expansion machine requiring only a small flow rate.

CHAPTER ELEVEN

Solid-Vapor Heat Pumps, Refrigerators, Heat Transformers, and Heat Storage Devices

In the preceding chapters, devices are described in which the absorbents were liquids. In Chapter Two, thermochemical reactions were shown also for solid absorbents. From a thermodynamic point of view, the processes of heat pumping, refrigeration, and heat transformation are not bound to the phase state of the absorbent. However, the chemical engineering, i.e., design of the heat exchangers and vessels, must be adapted to the properties of the phase state. Solids are difficult to pump; therefore, a solid bed has to be used, permitting mass and heat transfer. As a consequence of the solid bed, the processes of absorption and desorption are inherently discontinuous. To approach a quasi-continuous operation, at least two beds are required which are operated at a certain phase shift. On the other hand, the discontinuous mode of operation suggests that solid-vapor systems are used also for heat or cold storage. Therefore, in this chapter the principle of thermochemical heat storage will be discussed in more detail. Yet, it should be pointed out that thermochemical heat storage is also possible with liquid absorbents.

11.1 THE OPERATING PRINCIPLE OF SORPTION STORAGE DEVICES

Although thermochemical heat storage requires at least two containers (one for the absorbent and one for the condensed working fluid), higher energy densities can be obtained than are possible by employing heat of solidification or sensible heat. The heat of reaction per mole is about one order of magnitude larger than the heat of solidification at the same temperature level. The high energy density per unit volume is accomplished by storing the desorbed vapor as a condensed liquid or by absorbing it in a second absorber. In Figure 11.1a to d, the storage and discharge processes of single-stage heat (cold) storage cycles are shown superimposed on a $\ln(p)$,-1/T diagram. In Figure 11.2, the directions of heat flow and working fluid streams and the temperature levels of the two containers are indicated. For the sake of simplicity, it is initially assumed that all processes and reactions are reversible. In Figures 11.1a and 11.2a, heat is supplied at T_1 for desorption and heat is rejected at T_0 by condensation. For the discharge of heat, in Figures 11.1a' and 11.2a', heat is supplied at T_0 which evaporates the working fluid at the same pressure as previously obtained during the condensation process. Due to the absorption of the working fluid, useful heat is obtained at T_1. This operating mode returns the same amount of heat which was stored originally.

In the processes of Figures 11.1b, b' and 11.2b, b', the amount of useful heat is increased. This effect is obtained in addition to the storage process. The useful heat is delivered at a temperature level T_1, which is lower than the temperature level T_2 at which heat was supplied. However, the amount of useful heat is increased by the amount of heat absorbed from the ambient. One half of the useful heat is generated during the storage process; the other half is made available during discharge. Such a process may be employed whenever the temperature level of the heat source is higher than that of the heat required by the user.

Figures 11.1c, c' and 11.2c, c' show the inverse process. In this case, the useful heat is delivered on a higher temperature level, T_2, than the temperature level T_1 used to store heat.

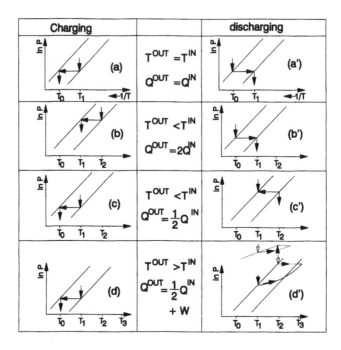

Figure 11.1 Pressures and temperatures of the three typical operating modes of a sorption storage unit (a,b,c).

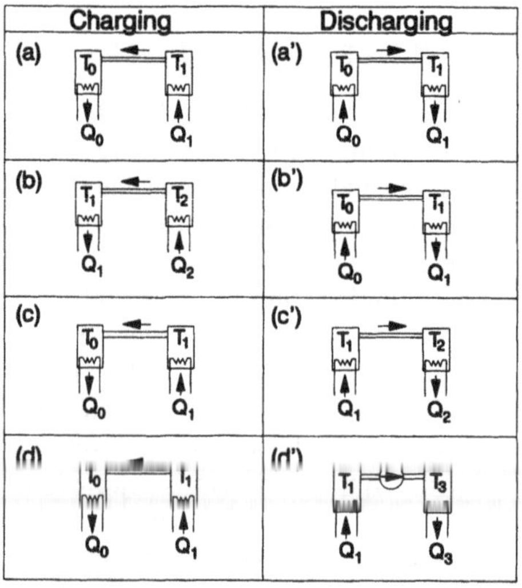

Figure 11.2 Heat flow and temperatures of the three typical operating modes of a sorption storage unit (a,b,c).

The amount of useful heat is smaller than the amount of heat required to drive the cycle. This type of process may be used to raise the temperature level of the stored heat, such as solar heat or waste heat. The processes in Figures 11.1 and 11.2 are heat conversion processes operating between four temperature levels. For the sake of simplicity, it was assumed in Figures 11.1b, c and 11.2b, c that the two intermediate temperature levels are identical. The changes in pressure, temperature, and volume of the working fluid during charging and discharging can be studied by inspection of Figure 4.7 (Chapter Four). Depending on the heat source used during the discharge process (which may occur at a different location), the temperature levels may be changed. In Figures 11.1d, d′ and 11.2d, d′, a heat storage process is shown which

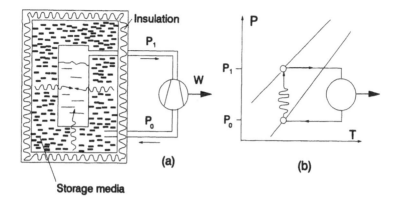

Figure 11.3 Thermochemical exergy storage unit for work.

achieves an even higher temperature increase to T_3 by integrating a compressor which increases the pressure of the working fluid evaporating at T_1. The compressor may work with a different working fluid represented by dashed lines. Heat is pumped from T_1 to T_2 to evaporate the working fluid of the storage cycle at T_2, generating heat at T_3.

In addition to the two operating processes already discussed for continuous absorption units, there exists the storage-only mode. The process is interrupted halfway through the cycle and then reversed to return to the original state. The three heat storage methods of Figures 11.1 and 11.2 may be characterized by their respective pressures during charging (P_1) and discharging (P_0):

(a) $P_0 = P_1$:   Plain storage process

(b) $P_0 < P_1$:   Heat storage with heat pump effect

(c) $P_0 > P_1$:   Heat storage with heat transformer effect

11.2 SORPTION UNIT FOR STORAGE OF AVAILABILITY

In reality, sorption storage units are not so much heat or cold storage devices as they are availability storage devices. Figures 11.1 and 11.2 show that any absorption of heat is accompanied by a rejection of heat. The availability of the two heat streams is different. Any change of the state of the storage device corresponds to a change of the availability of the system consisting of absorbent and working fluid.

Figure 11.3 shows how a sorption storage unit may absorb or deliver work. In this case work, W, is produced without the system exchanging heat with its environment. The inner container is filled with working fluid of saturation pressure P_1. The vapor is expanded in a turbine to pressure P_0 and is absorbed in the absorbent surrounding the inner container. The absorber heat is used to further evaporate the working fluid. By inverting the directions of all arrows, the sorption heat storage device is charged by the input of work. The working fluid is desorbed from the absorbent and condensed in the inner container while the heat of condensation is used to further desorb working fluid. A thermochemical storage unit can be used as a storage device for work in the same way as an electrochemical storage device. The availability content can be used for generating work or delivering heat at shifted temperature levels, thus providing heating and/or cooling capacity. In reality, the configuration of Figure 11.3a cannot be operated completely adiabatically because this would result in a change of pressures in dependence of the amount of work stored. Whether the temperature increases or decreases depends on the latent heat, the heat of absorption, and the efficiency of the turbine. In order to stabilize operating parameters, a small but important component has to be added, allowing supply or rejection of excess heat.

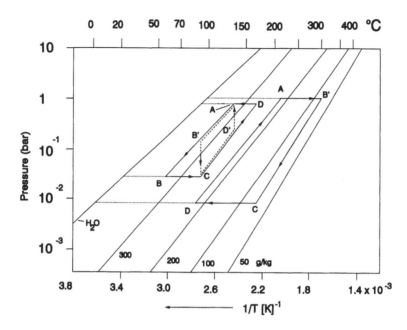

Figure 11.4 Vapor pressure curves of water over zeolite.

11.3 THE ZEOLITE HEAT OR COLD STORAGE DEVICE

The working pair water/zeolite may serve as an example for a heat pump or a heat transformer using a solid absorbent. Figure 11.4 shows typical vapor pressure curves for water which is adsorbed to a zeolite (alumina-silicate). In Figure 11.5, a detailed solution field for a Mg-A zeolite is shown. The far left vapor pressure curve is the one of pure water. In Figure 11.6, the main components of a zeolite (cold) heat storage unit are shown. The temperatures in Figure 11.6 correspond roughly to the heat pump cycle ABCD (clockwise) of Figure 11.4.

The process consists of four phases:

AB: Desorption phase, condenser provides useful heat.

BC: Cool-down phase, the sensible heat of the zeolite and of the heat exchangers is supplied as useful heat.

CD: Absorption phase, the heat of absorption, generated between C and D is available as useful heat. Refrigeration is available at the temperature of the evaporating water.

DA: Heating phase (no useful heat).

The heat required to preheat the containers, heat exchangers, and the zeolite is recuperated during the cool-down phase, but does not contribute directly to the heat pump effect. Consequently, the amount of sensible heat must be kept small compared to the heat of absorption. This may be accomplished by design and by choosing the cycle parameters such that a large difference in composition is obtained. The existence of the sensible heat is a disadvantage for the COP in heat pump operation, but can be exploited beneficially for storage purposes: The cool-down phase may be continued to at least the temperature of point D (Figure 11.4), without any absorption. When the absorption phase begins, the zeolite and the other components are heated by the heat of absorption. The absorption phase terminates at a temperature determined by the ratio of sensible heat to absorption heat, somewhere below point C. The heat stored as sensible heat can be used later as useful heat without requiring further absorption. This type of storage heat pump allows a time-wise decoupling. The useful heat is provided at

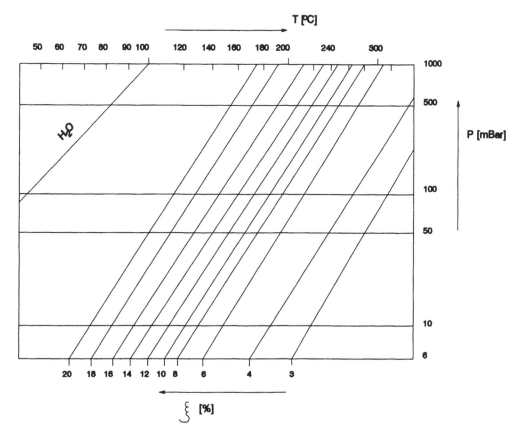

Figure 11.5   Vapor pressure curves of water and steam over a Mg-a zeolite.

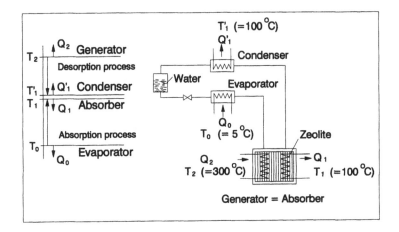

Figure 11.6   Sorption heat pump.

a certain time, while the driving heat and the heat of the surroundings are supplied at another time. A zeolite heat pump may be used, for example, to store low cost electric power supplied during nighttime or weekends. In a similar way, peaks in natural gas consumption can be lowered by charging the heat pump during times of low load, resulting in a more efficient use of the distribution net. Finally, such a system can be used as an air-conditioning unit, for

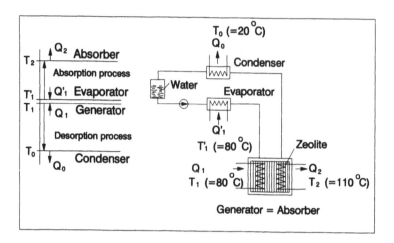

Figure 11.7 Sorption heat transformer.

example in automobiles, using waste heat and storing cooling capacity. The COP for a single-stage zeolite heat pump is in the range of 1.3 to 1.4.

In Figure 11.7, all essential components and temperature levels for the operation of a heat transformer, (ABCD or AB′CD′, respectively, counterclockwise operation according to Figure 11.4) are shown. In Figures 11.6 and 11.7, the evaporator is shown as an individual component separated from the absorber. To avoid pressure drops, the evaporator should be connected directly to the absorber with a large cross-section duct, as is done in LiBr/water units.

Other working pairs besides zeolites/H_2O are under test for solid absorbent air-conditioners and refrigerators. Examples are silicagel/water, Na_2S/H_2O, CaO/H_2O and carbon/NH_3, alkali and earth-alkali salts and NH_3, as well as metals/hydrogen systems. Using carbon and other absorbents for Ne, H_2, or He, refrigerators for rather low temperatures (0 K) have been proposed and are being tested.

Three major technical challenges have to be tackled when using solid absorbents:

1. The volume changes of the bed during the absorption and desorption process,

2. The mass transfer into and out of the absorbent bed,

3. The heat transfer into and out of the solid material, in between pellets and between the surfaces of the heat exchanger and the solid bed.

A promising method has recently been developed to cope with these challenges. The absorbent is imbedded into expanded graphite. The graphite is an elastic material absorbing the volume changes. It also provides channels for the working gas to enter and leave the absorbent, and, lastly, it enhances the heat transfer from the absorbent to the heat exchanger surface. In contrast to other heat-transfer enhancing materials, expanded graphite has a low specific heat capacity on a volume basis. Currently, quiet refrigerators are being developed on this basis for hotel room applications.

11.4 MULTISTAGE STORAGE DEVICES, STORAGE HEAT PUMPS, AND HEAT TRANSFORMERS

The diagrams of Figures 7.1 (Chapter Seven) and 9.1 (Chapter Nine) in conjunction with the tables listed in the Appendix provide all of the information required to construct multistage versions for heat pumps using solid-vapor systems. Storage can be added as an additional

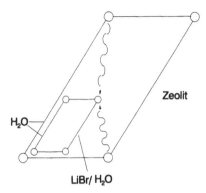

Figure 11.8   Triple-effect cascade.

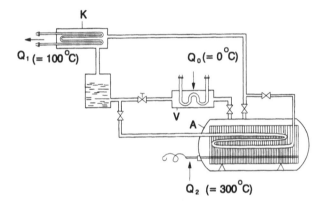

Figure 11.9 Zeolite high temperature heat pump and heat storage unit (A, desorber and absorber; K, condenser; V, evaporator).

feature. One or both stages may be operated discontinuously. Both stages may be coupled by heat transfer only (cascade) or by heat and mass exchange.

Figures 11.4 and 11.5 show that both the heat of absorption and the heat of condensation can be made available at such high temperature levels that both may be used as driving heat for a second stage or, for example, as the driving heat for a $H_2O/LiBr$ or $CH_3OH/LiBr/ZnBr$ heat pump. In Figure 11.8, this scheme is shown as a cascade according to Figure 7.1m. Figure 11.9 shows how the large absorption cycle of Figure 11.8, including the internal heat exchange, can be constructed. The condenser K is part of the generator of the low temperature cycle. During desorption in the high temperature cycle, the high pressure steam is condensed in K and stored in a receiver. During absorption, water vapor is produced in the evaporator, V, providing cooling. The vapor is absorbed in A, and the heat of absorption is transferred to K by evaporating water in the absorber cooling coil.

Assuming a COP of 1.4 for the zeolite heat pump and a COP of 0.70 for a single-stage LiBr/ H_2O - air-conditioner, the following COP is obtained for the combined system: 1.4 * 0.70 + 0.4 ≈ 1.4, a value approximately 20% larger than the heat ratio for the two-stage LiBr/H_2O machine. Heat storage with the zeolite stage may be an additional advantage. A heat ratio of 1.3 was demonstrated in a laboratory breadboard unit with a 20 kW zeolite cycle and a 10 kW LiBr/water cycle. This system was the first experimental example of a "triple-effect" unit.

Figures 11.10, 11.11, and 11.12 show how the two-stage configurations of Figure 7.1t,u,v can be operated with solid-vapor systems using essentially three containers. A and A' in Figure

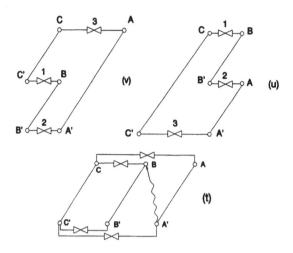

Figure 11.10   Schematic diagrams of two-stage storage heat pumps as seen in Figure 7.1t, u and v.

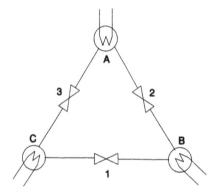

Figure 11.11   Principle of two-stage storage heat pumps according to Figures 7.1u through z.

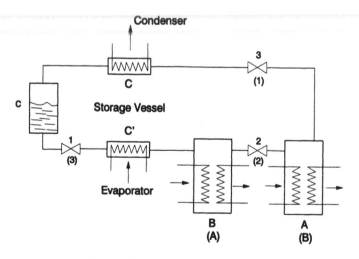

Figure 11.12   Two-stage storage heat pump of Figure 11.10v and u.

11.10 mark the identical container filled with absorbent, but for different pressure levels. On the high pressure level, A serves as a desorber; on the low pressure level, A′ serves as an absorber. The same holds for B and B′. The absorbent is made up of a different material than in A or of the same material at a different absorbent/working fluid concentration. C′ and C represent the storage container for liquid absorbent. The appropriate evaporator and condenser may be integrated or installed separately, as indicated in Figure 11.12.

The meaning of the two types of arrows (straight and wavy) is the following:

X → Y: Working fluid is flowing from X to Y.

X ⤳ Y: Heat is transferred from X to Y.

The storage and discharge processes of Figure 11.10u and v are conducted in three steps. The numbers in parenthesis represent the open connection in Figures 11.10 to 11.12. The processes marked by a star (*) produce cooling (or absorb heat from the surroundings).

	Version 1		**Version 2**
Configuration v	A → C (3) C′ → B * (1) B′ → A′* (2)	or	A → C (3) B′ → A′* (2) C′ → B * (1)
Configuration u	B → C (1) C′ → A′* (3) A → B′ (2)	or	B → C (1) A → B′ (2) C′ → A′* (3)

Configuration v achieves higher COPs than a single-stage device. Configuration u may be operated with the driving heat supplied at a temperature level which is relatively low compared to the temperature of the rejected heat. The heat exchangers shown for the components A, B, and C are used for heat rejection and for heat supply. In Figures 11.11 and 11.10, the three connections between the storage containers are numbered 1, 2, and 3. The six permutations of the relative locations of the three pressures within the connections (and therefore within the containers as well), represent all six configurations of Figure 7.1u to z, as has been shown in Figure 7.59 (Chapter Seven), column S_0. Thus, the configuration of Figure 11.11 allows obtaining of all operating modes (including heat transformers) of the configurations of Figure 7.1u to z. In fact, Figure 11.11 corresponds to Figure 7.48 (Chapter Seven). Since pairs of exchange units are combined within one container, A, B, and C, the total number of necessary connections is reduced from 12 to 3.

Even further modifications of the modes of operation as shown above for configuration v and u are possible. Because of the discontinuous operation, heat storage is possible in addition to heat pumping and heat transformation. Since processes may be started or terminated at any point, the number of possible modes of operation is greater for discontinuous machines than for continuous machines.

So far, the amount of working fluid generated by A and B are the same, requiring the same size for both exchange units (Figure 11.12). However, exchange unit B may be small compared to A, and still the effect of staging is maintained. Version 1 of configuration v may be modified as follows:

Version 3	A	→ C	
	1/2 C′	→ B	*
	B′	→ 1/2A′	*
	1/2 C′	→ B	*
	B′	→ 2/2A′	*

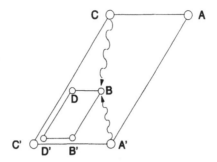

Figure 11.13 Triple-effect cascade.

In this case, exchange unit B is charged and discharged twice while A completes just one cycle. Generally, exchange unit B may cycle n-times per one cycle of A. B absorbs only the fraction 1/n of the working fluid of A per cycle and may, therefore, be small.

As a final example of the configurations of Figure 7.1a to r, version m is discussed, as shown in more detail in Figures 11.13 and 11.14. The following discussion is in principle also valid for units according to Figure 7.1s and t which are shown for solid vapor systems in Figure 11.15.

Since the condenser heat at C, Figure 11.13, is used for desorption purposes in B, condensation occurs simultaneously on two pressure levels. Therefore, two storage containers, C and D, Figure 11.14, for working fluid must be provided. Heat exchanger b in exchange unit B serves as a condenser for vapor generated from A, entering through line c. This establishes the internal heat exchange C $\rightsquigarrow$ B, Figure 11.14. With the help of heat exchangers a and b, the internal heat exchange A' $\rightsquigarrow$ B is established. For the configurations of Figure 11.15 in which only the heat of condensation or only the heat of absorption is transferred internally, the proper connecting lines in Figure 11.14 are omitted or rearranged.

For the configuration of Figure 11.14, the same considerations apply concerning the relative size of exchange units B and A. If B undergoes many (n) cycles during one cycle of exchange unit A, the size of B may be small. In Figure 11.15, examples for multistage configurations are given that combine liquid absorbent cycles with a cycle using CaO/H$_2$O as working pair. In Figure 11.15a through e, the liquid cycle is the LiBr/H$_2$O single-effect or double-effect unit, respectively. In Figure 11.15f through g, a triple-effect LiBr/H$_2$O is combined with the solid absorbent cycle. All of these cycles may be built as cascades or with water as the common working fluid in the form of an integrated multistage system. The liquid cycle in Figure 11.15h may be operated with ammonia/water (at higher pressures) as a double- or triple-effect cycle, including GAX or the branched GAX cycle. For all cycles, it holds that the flue gas leaving the high temperature generator can still be used in the lower temperature cycles.

The potential for the COP for cooling is rather large for the cycles in Figure 11.15 and may even exceed 2.0 for some of them.

11.5 QUASI-CONTINUOUS OPERATION WITH SOLID-VAPOR SYSTEMS

Single-stage solid-vapor systems can be operated quasi-continuously by duplicating the solid absorbent and by operating the two beds with a phase shift of half a cycle. Clearly, this method may be applied to any multistage cycle. Yet, for cycles like 11.10v or t, it is possible to produce cooling quasi-continuously just by separating the condenser and evaporators already shown in Figure 11.12, yet without the storage vessel. The modes of operation are as follows:

Cycle v   A → C   C' → B *
            B' → A'*
Cycle t   A → C   C' → B'*
            C' → A'*   A' $\rightsquigarrow$ B   B → C

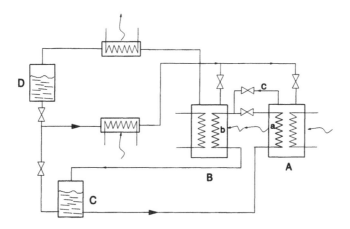

Figure 11.14 Two-stage storage heat pump of Figure 11.13.

If the cooling capacity is produced at the same rate in both steps, the heat output varies by about a factor of two between the first step and the second step of the process. If the heat output rate is kept constant, the cooling capacity is different for the two steps.

The cycling of the absorbent beds leads to large losses in efficiency. For units with liquid absorbents, these losses are reduced by the solution heat exchanger. For solid absorbents, several methods for internal heat exchange have been proposed: the intermediate connection of the two vessels before the reversal of the operation. The working fluid carries heat by desorption and absorption from the high pressure vessel to the low pressure vessel. A heat transfer fluid may be circulated through both vessels serving as the primary energy source, but also providing internal heat exchange. By this method not only specific heat is exchanged, but also high temperature of absorption can be used for desorption, thus saving primary energy and improving the COP. This principle of absorber-generator heat exchange is, from a thermodynamic point of view, identical to the method used in the double-effect cycles. Internal heat exchange by a heat transfer fluid is particularly effective if a zone-type absorption and desorption process is achieved.

11.6 SOLID-VAPOR SYSTEMS WITH COMPRESSION/EXPANSION MACHINES

Generally, all configurations of Figure 9.1o to t can be constructed according to Figure 11.16 which contains two exchange units, C and B, and a compressor K. Figure 9.25, column S_0, indicates that the configurations are distinguished only by the relative pressure levels prevailing in connections 1, 2, and 3 and the compression/expansion machine.

Figure 11.17 shows the implementation of the configuration of Figure 9.1o (Chapter Nine) with a solid-vapor system using a compressor. To produce cooling, the compressor first charges the system by compressing working fluid vapor from B which is condensed in C. The heat of condensation is used in B for further desorption. During discharge, the liquid working fluid evaporates in C', producing cooling capacity, and is absorbed in B'. In order to balance the internal heat exchange during compressor operation, a small exchange unit, E, is introduced. Figure 11.18 shows a more detailed process scheme of the configuration of Figure 11.17. The internal heat exchange occurs between C and B. During the storage phase, the condensation heat of C is used for desorption in B. The storage unit stores the availability supplied by the compressor. Device E plays an important role in maintaining a stationary operation. The condensate from C is expanded to the pressure of E. The flash vapor is supplied

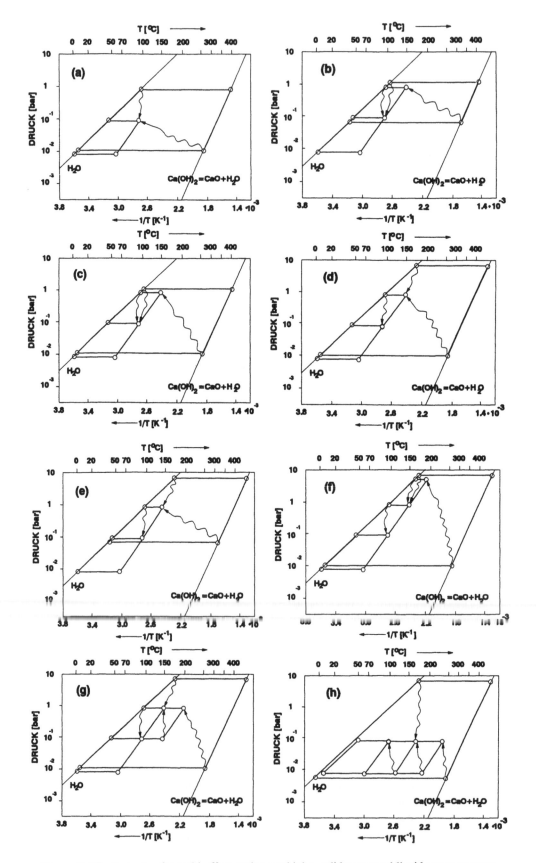

Figure 11.15 Examples for multi-effect cycles combining solid-vapor and liquid vapor systems.

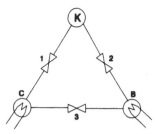

Figure 11.16 Principle of a storage heat pump including a compressor according to Figure 9.1o to t.

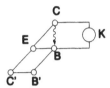

Figure 11.17 Process scheme of a storage heat pump.

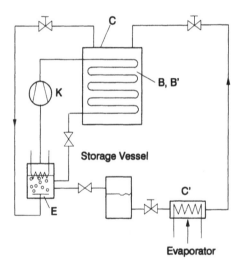

Evaporator

Figure 11.18 Storage heat pump. Charging: v_1, open; v_2, v_3, closed; K, in operation. Discharging: V_1, closed; V_2, V_3, open; K, not in operation.

to the suction side of K. In addition, the superheated vapor desorbed in B is cooled in E before entering the compressor. Both effects cause an increase in COP. The most important task of E, however, is the following: Since the condensation heat and the desorber heat per unit of working fluid are not the same, the difference, usually a small portion of heat, has to be supplied or rejected by E in order to obtain a stationary operation.

REFERENCES

1. Internal Seminar of Thermochemical Energy Storage, Stockholm, January 7-9, 1980 ed. Gunnar Wettemark. Swedish Board of Building Research.

2. W. Pauer, Energie Speicherung, Verlag Theodor Steinkopf, Dresden, 1928.

3. F.R. Kalhammer, Energy storage systems, *Sci. Am.*, 241, No. 6, p. 42 (1979).

4.   G. Alefeld, Basic Physical and Chemical Processes for the Storage of Heat, Proc. Symp. on Electrode Materials and Processes for Energy Conversion and Storage. Var. 1977, Philadelphia, J.E.E. McIntyner et al., Ed., Proc. of the Electro-Chemical Society, Vol. 77-6.

Solid Absorbents:

5. Proceedings of the Symposium: Solid sorption refrigeration, Paris, France, November 18 - 20, 1992, Editor F. Meunier et al. ISBN 2903 633, Int. Inst. of Refrigeration, Paris.

Graphite Heat Exchangers:

6. B. Spinner, Proceedings of the Symposium: Solid sorption refrigeration, Paris, France, November 18 - 20, 1992, Editor F. Meunier et al. ISBN 2903 633, Int. Inst. of Refrigeration, Paris.

7. M. Coulon, R. Faron, M. Besson, Patent FR 2610088-A1 (1987) and US Patent 4,852,645 (1989).

Triple-Effect Cycle:

8. F. Ziegler, F. Brandl, J. Volkl, G. Alefeld, A cascading two-stage sorption chiller system consisting of a water-zeolite high-temperature stage and a water-lithiumbromide low-temperature stage, Proc. Absorption Heat Pump Congress, Paris, March 20–22, EUR 10007 EN; 1985; 231–238.

9. G. Alefeld, F. Brandl, J. Volkl, F. Ziegler, Advanced cycles for the use of solar energy; Int. Symposium on Thermal Applications of Solar Energy, April 1985, Hakone, Japan.

Thermal Wave:

10. S.V. Shelton, D.J. Miles, Solid/vapor thermally driven heat pump development, Int. Workshop on Research Activities on Advanced Heat Pumps, Graz, Austria, 1986, dbv-Verlag, Graz.

11. D.I. Tchernev, Regenerative zeolite heat pump in: *Zeolites: Facts, Figures, Future.* Editors P.A. Jacobs, R.A. van Santen; Elsevier, Amsterdam, The Netherlands, 1989, 519–527.

CHAPTER TWELVE

Open Heat Transformation and Storage Systems

Most of the absorption and vapor-compression systems, whether they operate continuously or discontinuously, are closed machines. For the low pressure system LiBr/H$_2$0, an automatic purging device is required. Open absorption systems are mainly used for the dehumidification of air (e.g., with LiCl solutions as absorbers) or for drying of air (e.g., with zeolites as absorbers in compressed air systems). Other applications are the separation of gas streams by molecular sieves.

12.1 OPEN HEAT CONVERSION SYSTEM FOR STEAM GENERATION

In Figure 12.1, A represents a generator, B an absorber, C a condenser, and D an evaporator. Between A and B, an absorber solution such as H$_2$O/LiBr, H$_2$O/LiCL, H$_2$O/H$_2$SO$_4$, or H$_2$O/ NaOH is circulated. The condenser heat of C serves as a generator heat in A; the absorption heat in B serves as evaporator heat in D. The configuration in Figure 12.1 converts high pressure stream at P$_2$ and low pressure steam at P$_0$ into steam of intermediate pressure levels P$_1$ and P$'_1$. In the configuration of Figure 12.2, all heat and fluid steams are reversed. Steam at the intermediate pressure levels P$_1$ and P$'_1$ is converted into high pressure stream at P$_2$ and low pressure steam at P$_0$. With water/zeolite as working fluid, steam of 1 bar can be upgraded to steam of 3 to 5 bar. The low pressure steam is transferred to the ambient using air as carrier fluid. For continuous operation, two zeolite beds operating with a phase shift of half a cycle are required in this case.

12.2 OPEN STORAGE SYSTEMS

When moist air is passed through a dry, water vapor absorbing bed like silicagel, zeolite, alumina, or wood, a certain amount of water vapor is absorbed, depending on the temperature of the bed and the partial pressure of the water vapor. The carrier fluid, air, serves simultaneously to remove the heat of absorption. When, on the other hand, dry warm air is passed through a bed charged with water vapor, the bed will be dried and the air cooled. By a combination of these processes, the modes of operation as described in Chapter Eleven, including the multistage heat conversion and storage modes, can be implemented. Air serves as transfer fluid for water vapor and simultaneously for heat and/or refrigeration.

Open storage systems have the advantage in that they do not have to be air tight. In contrast to closed systems, they require a large pumping power to establish the required air-flow rates. Also, the reaction speed is considerably smaller than in vacuum systems, due to the diffusion of steam through air. Reaction zones become more diffuse. Open systems also suffer from contamination and decomposition of the storage medium by impurities in the air. Filters are required. Depending on the application, either open or closed systems are promising. Currently, open systems are considered to assist conventional air-conditioners in the dehumidification process.

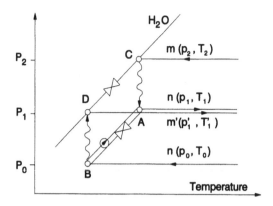

Figure 12.1   Open absorption heat pump.

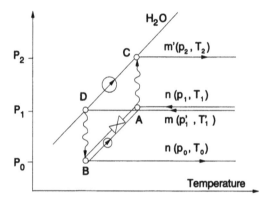

Figure 12.2   Open absorption heat transformer.

Problems

1. Heat Transformation

Three identical bodies of steel with a mass of m = 10kg and a specific heat capacity of c = 0.5kJ/kgK have the following temperatures: $T_1 = T_2 = 300°C$ and $T_3 = 100°C$. There is no heat exchange with the surroundings.

(a) Now the three bodies are brought into thermal contact. What is the temperature of the three bodies after thermal equilibrium is reached? How much is the increase in entropy?

(b) What is the maximum work that can be produced by the heat stored in the three bodies? What is the equilibrium temperature in this case, and how much does the entropy increase?

(c) What is the highest or lowest temperature that can be reached by one of the three bodies when heat or work is exchanged only among the three bodies? How much does the entropy increase? Describe three processes to obtain the highest or lowest temperature.

(d) How does the outcome of (c) change with increasing irreversibility of the processes. Which of the three processes of (c) do you assume to be the most beneficial?

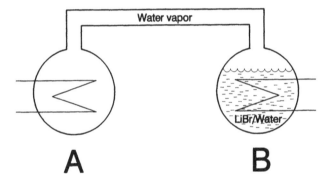

Figure P.1.

2. Heat Transformation with Absorption Processes: Heat Storage

Figure P.1 shows a pair of containers, A and B. Container B contains an aqueous solution of lithium bromide at 90°C with a mass fraction x of 60% salt. Container A and the connecting pipe are filled with steam only. The specific heat capacity of steam is c″ = 1.8 kJ/kgK, of water it is c′ = 4.2 kJ/kgK, and that of the solution is c_{sol} = 2.2 kJ/kgK. All other data can be read from Figures 2.15 and 2.1, Chapter Two. The heat capacity of the containers may be ignored.

(a) What is the pressure in A?

(b) From container A, heat is rejected.

(i) At what temperature can the first droplets of water be observed?

(ii) By what means can the pressure be kept constant?

(iii) While the water evaporates from B, the mass fraction of salt in B increases. Calculate the change Δx in mass fraction of salt as a function of the amount of steam produced. Use as variable f = (mass of solution initially)/(mass of steam produced).

(iv) As the mass fraction of salt increases, the temperature of the solution increases as well (pressure is kept constant). Specify the temperature change as a function of f, using Figure 2.15. What is the temperature T_s of the solution when the mass fraction is changed by 0.05? What is the amount f of solution required to produce 1 kg of steam?

(v) What is the latent heat r of water evaporating from liquid water and from the solution respectively? (use Clapeyron's equation)

(vi) How much heat per kilogram of steam is rejected in A and has to be supplied to B?

(c) To return the solution to the initial state, heat is added to container A, evaporating the water. Container B is cooled to absorb steam into the solution. Again, the pressure is kept constant.

(i) At a mass fraction of 0.6, what are now the amounts of heat exchanged in A and B?

(ii) During this process, heat is added to A at a temperature lower than the temperature in B, while more heat is removed from B at the higher temperature than was added to A. How is this consistent with the First and Second Laws?

(d) The process of absorption and desorption may be used to store heat.

(i) What fraction $\eta = q_A(evap)/q_A(const)$ of the heat input can be recovered?

(ii) Why is the storage process not fully reversible?

3. Heat Transformation with Absorption Processes: Heat Pump

During the process described in 2(c) above, the pair of containers may operate at a different pressure level compared to 2(b). To obtain a lower pressure, the temperature in both containers is reduced. To obtain a higher pressure, the temperature of both containers is increased. The temperature change is assumed to occur without the exchange of steam between A and B. The pressure in A is always equal to that in B.

Now two cases are considered:

When the pressure level is chosen to be lower than in 2(c), it is assumed here that at the end of the absorption process the temperature of the solution is the same as the temperature of container A in 2(c). When the pressure level is chosen to be higher than in 2(c), it is assumed that the water evaporates at the same temperature as that of container B at the end of the absorption process in 2(c). For the following calculations, the temperature dependence of the latent heat is ignored:

(a) (i) Discuss the processes that occur during the change of the pressure level.

(ii) Plot the processes during a pressure change in the $\ln(P)$ vs. $-1/T$ diagram, specifying temperatures and pressures.

(iii) Calculate the amounts of heat exchanged during heating or cooling of the liquid in the containers.

(b) When the new pressure level is reached, steam is produced in A and absorbed in B. Which case represents the heat pump and which represents the heat transformer?

(c) The pressure level in the pair of containers is adjusted to the original value.

(i) Calculate the amounts of heat exchanged in this process.

(ii) Define for the two cycles a coefficient of performance (useful heat)/(expended heat) and calculate its value.

4. Absorption Chillers

An absorption heat pump using the working fluid pair water/lithium bromide is evaluated. The evaporation temperature for the water is $T_0 = 5°C$. The condensing and the lowest absorber temperatures are $T_1 = 35°C$. The change of concentration in absorber and generator is 5%.

Preparation:

(a) Sketch a process scheme of the unit.

(b) Mark the pressures and temperatures of the main components: evaporator E, absorber A, generator G, and condenser C on the $\ln(P),(-1/T)$-diagram.

(c) What is the minimum temperature requirement for the driving heat T_2?

(d) What is the maximum heat ratio, COP_{Carnot}, q_E/q_G?

(e) Determine the ratio of the heat of solution l to the latent heat of evaporation r at the average absorber concentration.

(f) Calculate the amount of water-rich absorbent f that has to be circulated to produce 1 kg of vapor in the generator by employing mass balances for solution and water.

$$f = \frac{\text{mass of water-rich solution}}{\text{mass of steam generated}}$$

Calculation of the amounts of heat exchanged:
The calculation is based on 1 kg of water circulating in the system. The following properties are assumed to be temperature independent. Specific heat capacity of water $c' = 4.2$ kJ/kgK, specific heat capacity of steam $c'' = 1.8$ kJ/kgK, specific heat capacity of solution $c_{sol} = 2.2$ kJ/kg, density of the solution $= 1700$ kg/m³. The latent heat of water is 2460 kJ/kg at 5°C and 2420 kJ/kg at 35°C.

(g) What is the pump work, assuming reversible operation?

(h) Calculate the evaporator heat and condenser heat.

(i) The heat input to the generator is used to preheat the solution to the saturation temperature, to desorb steam, and to further heat the remaining solution to the generator exit temperature. Calculate the generator heat requirement.

(j) What is the amount of heat rejected in the absorber? Check the energy balance!

Discussion:

(k) What is the ratio of cooling capacity to driving heat?

(l) Usually a counter-flow heat exchanger is used for heat exchange between the solution streams. Should this heat exchanger be on the high pressure side or on the low pressure side?

(m) What is the maximum amount of heat that can be transferred? Is this a reversible process?

(n) What is the heat ratio (COP) in this case? What is the ratio of heat ratio over Carnot efficiency?

5. Superposition

Figure P.2 shows a two-stage absorption refrigeration system with improved efficiency (double-effect system). The arrows at the corners indicate the direction of heat exchange. The wavy line indicates a completely internal heat exchange within the cycle.

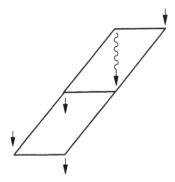

Figure P.2.

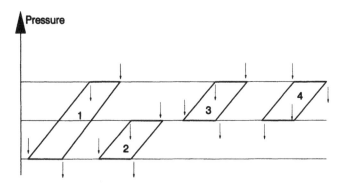

Figure P.3.

(a) Describe the phase changes in the six heat exchangers.

(b) Two-stage units can be represented as the superposition of two single-stage heat pump or heat transformer cycles. Find all combinations of the basic cycles 1 through 4, Figure P.3, that can be combined to the double-effect heat pump.

(c) The heat ratio θ of the refrigeration units is the ratio of cooling capacity to driving heat. It is assumed that $\theta_1 = 0.6$, $\theta_2 = 0.8$, and $\theta_3 = 0.7$. The heat ratio for the heat transformer, cycle 4 in Figure P.3, is the ratio of useful heat to driving heat and assumed to be $\theta_4 = 0.5$. Calculate the values for the heat ratios of the double-effect refrigeration unit based on the various options for combining the basic cycles.

(d) Which combination of the basic cycles best represents the heat ratio of the two-stage cycle? (Compare the irreversibilities!)

6. **Superposition**
   Figure P.4 shows three two-stage absorption refrigeration machines with increased efficiency.

   (a) Two-stage units can be seen as superpositions of two basic heat pump or heat transformer cycles which are coupled by heat exchange. Find the three options that exist to combine basic cycles to each of the configurations of Figure P.4.

   (b) The ratio of cooling capacity to driving heat for a single-effect unit is assumed to be $\theta_1 = 0.7$. The ratio of useful heat to driving heat for the transformer cycle is assumed to be $\theta_2 = 0.47$. Calculate the ratio of cooling capacity to driving heat for the three configurations in Figure P.4, using superposition.

(c) Discuss the advantages and disadvantages of the three configurations with respect to the working fluid pair used.

(d) Find configurations for which the ratio of cooling capacity to driving heat exceeds 1.6. (There may be more than two basic cycles required.)

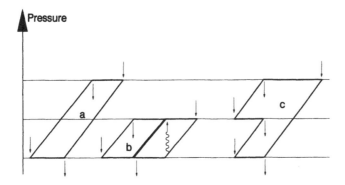

Figure P.4.

7. Entropy and Work

This problem clarifies the difference between work and heat with regard to entropy. Calculate and discuss the exchange of work, heat, and entropy in the following examples:

(a) A mass is lifted.

(b) The mass drops to the original level.

(c) A fluid is moved very slowly from a pipe into a container. Temperature and pressure are kept constant.

(d) The pressure of an incompressible fluid is increased adiabatically.

(e) An incompressible fluid is pumped from a pipe of low pressure into a pipe of high pressure.

(f) The temperature of an ideal gas is increased by adding heat at constant volume.

(g) The temperature of an ideal gas in a container is increased by adiabatic compression.

(h) An ideal gas is pumped from a pipe of low pressure into a pipe of high pressure.

(i) Assume that the same amount of energy is added to the ideal gas in (f) and (g) as heat or work, respectively. What is the difference between the final states? Explain the result by using arguments based on statistical thermodynamics.

8. Entropy Analysis for a Vapor-Compression Refrigeration System

The temperature in a space is maintained at $T_N = -10°C$. The temperature of the environment to which the waste heat is rejected is $T_R = 20°C$. To allow for heat transfer, the evaporator saturation temperature is $T_V = -15°C$, and the condenser saturation temperature is $T_K = 40°C$. The isentropic efficiency of the compressor is $g_k = 0.7$.

(a) Sketch the cycle in a T,s-diagram.

(b) What is the COP of the process using R22 data?

(c) Derive an energy balance and an entropy balance for this process. Based on these two equations, derive an expression for the work W as a function of the external temperatures, for the cooling capacity Q, and for the entropy generation of the system.

(d) Calculate the entropy production for each individual component in a cycle with 1 kW cooling capacity.

(e) Calculate the COP using the result of (d) and compare with the result of (b).

(f) What is the value of the COP when the area of the condenser is infinitely large?

9. Vapor-Compression Refrigeration Unit

A vapor-compression cycle that uses R22 as the refrigerant is considered. The following values are given:

Evaporator at $T_0 = -20$, $P_0 = 2.5$ bar, latent heat = 221 kJ/kg,

Condenser at $T_1 = 50$, $P_1 = 19.3$ bar, latent heat = 154 kJ/kg,

Specific heat capacity of the liquid $c_p' = 1.2$ kJ/kgK and of the vapor $c_p'' = 1.0$ kJ/kgK.

The compressor has an isentropic efficiency of $g_c = 0.7$. (Assume temperature independence of all properties.)

(a) Sketch the cycle in a ln(p),h-diagram and in a T,s-diagram.

(b) Calculate the enthalpy and entropy values at dew and at boiling lines for P_1 and P_0 (assume enthalpy = 0 and entropy = 0 for saturated liquid at P_0; ignore v_cP).

(c) Specify the initial and final states for the isentropic compression and realistic compression. What is the amount of work in either case?

(d) What processes occur during the throttling process? What is the vapor quality x after throttling?

(e) Calculate the cooling capacity and condenser heat per kilogram of refrigerant.

(f) Calculate the COP for the unit for heating and cooling.

(g) Define a Carnot efficiency for this process and justify your assumptions. What is the ratio of actual COP divided by the chosen Carnot efficiency?

10. Vapor-Compression Refrigeration System

A vapor-compression refrigeration system uses ammonia as the refrigerant (Figure P.5). The cooling capacity is 100 kW. The condenser operates at 30°C, the evaporator at 10°C. The isentropic efficiency of the compressors is 0.7.

Case A: Compression and expansion are single-stage processes.

Case B: Compression and expansion are two-stage processes. The intermediate pressure level is 3 bar. The discharge gas of the low pressure compressor (5, Figure P.5) is mixed with the condensate (8) that was expanded to the same pressure level. The vapor from the mixing vessel is the suction gas (6) for the high pressure compressor. The liquid leaving the mixing vessel is expanded to the low pressure level.

(a) Sketch the processes in the ln(P),h-diagram for ammonia.

(b) What are the suction volume flow rates for the compressors?

(c) Calculate the COP for both cycles.

(d) Why is the COP of the two-stage system higher?

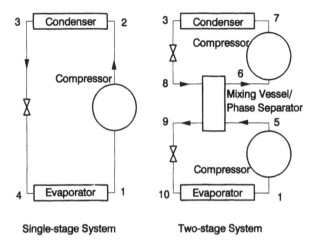

Figure P.5.

11. Comparison of Absorption and Vapor-Compression Refrigeration Units

Some basic differences between absorption and vapor-compression machines can be studied as follows: In both cases, a unit is considered that produces chilled water at $T_o = 5°C$ and waste heat at $T_1 = 35°C$, which is rejected by a cooling tower. The ratio of latent heat for evaporation from refrigerant over absorbent is $r_r/r_a = 0.91$.

(a) What is the COP of the vapor-compression unit, assuming reversible operation?

(b) What is the temperature T_2 of the driving heat for the absorption unit?

(c) What is the COP of the absorption unit, assuming reversible operation?

(d) Estimate the real COP for both units based on material presented in the text.

In winter, the temperature of the cooling tower can be reduced. Discuss qualitatively:

(e) How does the reversible and realistic COP of the vapor compression system change?

(f) How does the reversible and realistic COP of the absorption system change?

(g) What options become available for the absorption unit due to the lowered cooling tower temperature?

12. Gas Cycle Refrigeration System

A gas cycle refrigeration system is analyzed that may be suitable for car air-conditioning. The cycle is the reversed Joule Cycle using air (considered to be an ideal gas) as the working fluid. Heat rejection to the surroundings occurs at temperatures above $T_3 = 50°C$; the cooling capacity is provided at temperatures below $T_1 = 0°C$. Additional values are specified as follows:

State of the surrounding air: $T_0 = 40°C$, $P_0 = 1$ bar,

Temperature of the cold air entering the conditioned space: $T_s = 10°C$,

Specific heat capacity of air: $c_p = 1$ kJ/kg,

Isentropic exponent of air: $= 1.4$,

Isentropic efficiency for compression: $\eta_1 = 0.7$,

Isentropic efficiency for expansion: $\eta_2 = 0.9$.

Part 1: A closed cycle is implemented.

(a) Sketch the cycle in a T,s-diagram.

(b) What is the minimum value of the pressure ratio required to obtain the given operating temperatures?

(c) Calculate the COP of the process. What is the COP value for reversible compression and expansion? Which pressure ratio yields the optimum COP?

Part 2: The Joule Cycle can also be operated as an open system. Surrounding air is compressed adiabatically, cooled by using surrounding air, and then expanded adiabatically to the pressure level of the surroundings. The cold air is supplied to the conditioned space.

(d) Sketch the process in the T,s-diagram.

(e) What is the value for the pressure ratio for this process?

(f) Calculate the efficiency of the process.

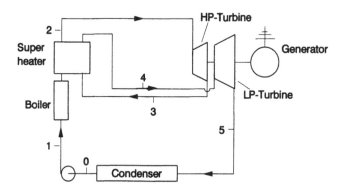

Figure P.6.

3. **Rankine Cycle in Various Diagrams**
   A Rankine Cycle is considered as described in Figure P.6. It is assumed that the turbines are reversible and that the work input to the pump is negligible compared to the work produced in the turbines. The working fluid is water. Steam is considered to be an ideal gas. Sketch the cycle in the following thermodynamic diagrams:

   $\ln(P),(-1/T)$
   $\ln(P),\ln(v)$ $T,\ln(v)$
   $\ln(P),h$ T,h $\ln(v),h$
   $\ln(P),s$ T/s $\ln(v),s$ h,s

   Also indicate the locations of the two-phase range and the critical point. Discuss the meaning of the slope of the various lines of constant properties in the diagrams.

14. **Steam Power Plant**

Consider the Rankine Cycle of Figure P.6. The following information is specified:
Evaporation, $T_e = 350°C$; 165 bar; maximum temperature of the superheat, $T_2 = T_4 = 550°C$;
condensation, 30°C; average specific heat of steam, $c'' = 2$ kJ/kgK; average specific heat of
water, $c'' = 4$ kJ/kgK; latent heat at 350°C, 1000 kJ/kg (now regenerative heating, reversible
components).

(a) Derive equations for the entropic average temperature T for heat supplied to the cycle
with and without superheat at the intermediate pressure level.

(b) Show qualitatively that an intermediate pressure level exists which yields maximum
COP.

(c) Calculate the optimum temperature T_3 and the intermediate pressure level. Assume the
steam to be an ideal gas.

(d) Calculate the efficiencies of the optimized cycle and of the cycle without intermediate
superheat.

15. **Internal Combustion Engine-Driven Heat Pump**

Figure P.7 shows a schematic diagram representing the energy streams in a diesel engine
driven vapor-compression heat pump.

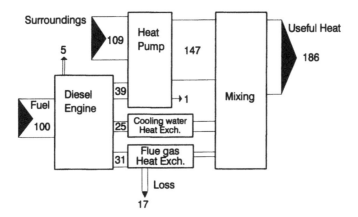

Figure P.7.

(a) Show a component diagram of the unit.

(b) What is the COP = (useful heat)/(work input) for the heat pump?

(c) Compare the efficiency of the system of Figure P.7 based on primary energy to that
of an electric-driven heat pump. The efficiency of the power plant is assumed to be
35%. Sketch a schematic diagram representing the energy streams.

(d) How much fuel is saved compared to a furnace with a burner efficiency of 85%?

(e) Is the heat pump still useful when the COP drops to half its original value due to an
increased temperature lift?

16. **Cogeneration of Heat and Power**

Two options for the use of stationary internal combustion engines $\eta_m = 0.4$ are compared.

A. The entire work produced by the engine is used to operate a vapor compression heat pump with a COP for heating of $\varepsilon = 2.5$. Also, 80% of the waste heat of the engine is reclaimed and used for heating ($\eta_B = 0.80$). It is assumed that such a combination of engine and heat pump replaces a conventional furnace with a burner efficiency of 0.80.

B. The engine is now designed such that 80% of the waste heat is reclaimed and used for heating, while the work is used for the production of electricity. The engine is used partially to replace electricity from conventional power plants ($\eta = 0.35$) and heat from furnaces with a burner efficiency of 0.80.

(a) Calculate for A and B the amount of energy saved, ΔE, based on the power output of the engine P: E/P.

(b) Which option provides the larger energy savings? Discuss the parameters involved and their influence on the result.

(c) Clarify the results by comparing energy streams!

17. Power-Refrigeration Cogeneration

Three options to produce work and refrigeration simultaneously are being considered.

A. Work is produced by a power plant with a thermal efficiency of $\eta = 0.38$. A portion of the work is used to operate a refrigeration unit with a $COP_r = 4.6$.

B. A cogeneration plant produces work and heat. The heat will be used entirely to drive an absorption heat pump. The ratio of heat to power can be varied to a certain extent. The cogeneration plant can be seen as a combination of two power plants. The first produces power at an efficiency of η_2. The waste heat leaving the first plant can be used in any proportion to either drive the absorption system or the second power plant which has an efficiency of η_2. If no heat is extracted, the overall efficiency for power generation is equal to that of option A. There are two options for the absorption system:

1. A single-effect absorption system is used with a $COP_{a1} = 0.8$. The driving heat has to be supplied at 90°C and $\eta_1 = 0.14$.

2. A double-effect absorption system is used with a $COP_{a2} = 1.35$. Now the driving heat has to be supplied at 150°C and $\eta_1 = 0.25$.

The efficiency for the distribution of heat η_H and electricity η_P is 0.95.

(a) Sketch a process schematic diagram for each of the options and indicate all components and energy streams.

(b) Find an expression for the maximum efficiency of the cogeneration plant (no extraction of heat) as a function of η_1 and η_2.

(c) Calculate the value for the expression W/P as a function of R/P with W representing the amount of primary heat, P the net work, and R the cooling.

(d) What is the maximum value of the ratio of cooling capacity to net work for the single-effect and double-effect absorption systems?

(e) Plot the results of (b) and discuss.

18. Analysis of Heat Exchanger Networks with the Pinch Method

The analysis of a simple production facility yields the following heat streams:

Stream No.	Type	Temp. Range (°C)	Capacity (MW)
1	Heating	30 to 80	5.0
2	Evaporation	80	9.0
3	Heating	130 to 240	5.0
4	Heating	185 to 195	4.5
5	Cooling	205 to 170	8.5
6	Condensation	100	9.0
7	Cooling	100 to 30	9.5

Assume that the heat quantities are distributed linearly over the respective temperature intervals.

(a) What is the amount of the entire heating load and entire cooling load? Estimate how much heat can be recovered by heat exchange.

(b) Determine for each temperature interval the heating load (Hint: The temperature interval between 130°C to 240°C has to be broken up into three temperature intervals). Plot these heat quantities as a function of temperature. Shift the sections of this plot related to each individual temperature interval along the heat axis such that the beginning of a new segment is placed at the end of the previous segment. Make the same plot for the heat quantities to be rejected.

(c) Shift the two curves along the heat axis such that the maximum amount of heat can be exchanged between the cooled and heated streams. Read from the plot the amount of heat recuperated and the remaining heating and cooling requirement. Compare the result with your estimate and (a).

(d) How much heat can be recuperated within the facility when the smallest temperature difference in the total heat exchange network (pinch point) is 10 K? What is the amount of the remaining heating and cooling requirement?

(e) Discuss by what considerations an economically optimum temperature difference at the pinch point can be derived.

(f) Sketch a schematic diagram that demonstrates the heat exchanger network. Discuss which problems may be encountered when such a network is implemented.

19. Solid-Vapor System

What modifications are required to the cycle in 11.10u to operate it quasi-continuously for cooling or for heating? Operate cycles 11.10u,v quasi-continuously as heat transformers. Operate cycle 11.10t quasi-continuously for cooling or for heating in the double-lift mode (heat exchange between C and B'). Operate cycle 11.10t quasi-continuously as a heat transformer in the double-effect or double-lift mode.

Problem Solutions

1. Heat Transformation

(a) $T_a = (T_1 + T_2 + T_3)/3 = 233.3°C$

$\Delta S = mc \ln (T_a^3/(T_1 T_2 T_3)) = 0.29$ kJ/K

(b) $\Delta S = 0$

$$T_b = 3\sqrt{T_1 T_2 T_3} = 223.6°C$$

$W = 3 mc(T_a - T_b) = 145.5$ kJ

(c) $\Delta S = 0$

At the end of the processes, two bodies must be at equal temperature. Therefore, the starting point with $T_1 = T_2 = 300°C$ and $T_3 = 100°C$ represents the solution with the lowest temperature. The highest temperature can be reached as follows:

power station between body 1 and 3:

(1) heat pump between body 1 and 2; highest temperature: body 2.

(2) heat pump between body 3 and 2; highest temperature: body 2.

(3) heat pump between body 2 and 1; highest temperature: body 1.

highest temperature $T_c = 380°C$ from:

$$(3T_a - T_c)^2 T_c = 4T_b^3$$

(d) T_c decreases;

for case 1, the smallest amounts of heat and work have to be exchanged.

2. Heat Transformation with Absorption Processes: Heat Storage

(a) 90 mbar

(b) (i) 44°C

(ii) By heating of vessel B

(iii) $\Delta x = x/(f - 1)$

(iv) $\Delta T/\Delta x = 2.25$ [K/% salt] for $x = 60\%$

$\Delta T = 135/(f - 1)$ [K]

$T_s = 101°C$, $f = 13$

(v) $r_{H_2O} = 2400$ kJ/kg; $r_{sol} = 2660$ kJ/kg

(vi) $q_A = r_{H2O} + c''(T_B - T_A) = 2480$ kJ/kg

$q_B = r_{sol} + c_{sol} \Delta T/\Delta x = 2960$ kJ/kg

(c) (i) $q_A = r_{H2O} = 2400$ kJ/kg

$q_B = r_{sol} + c_{sol} \Delta T/\Delta x - c''(T_B - T_A) = 2870$ kJ/kg

(d) (i) 97%

   (ii) specific heat of vapor

3. Heat Transformation with Absorption Processes: Heat Pump

(a) (i) $p = 8$ mbar: $T_A = 4°C$ $T_B = 44°C$

     $p = 700$ mbar: $T_A = 90°C$ $T_B = 144°C$

  (ii) pressure reduction:

$$q_A = c'(44 - 4) = 170 \text{ kJ/kg}$$

$$q_B = (f - 1)c_{sol}(102 - 55) = 1240 \text{ kJ/kg}$$

  (iii) pressure increase:

$$q_A = c'(90 - 44) = 190 \text{ kJ/kg}$$

$$q_B = (f - 1)c_{sol}(156 - 102) = 1430 \text{ kJ/kg}$$

(b) pressure reduction, heat pump

   pressure increase, heat transformation

(c) (i) $q_A = 0$ (empty container!)

$$q_B = f{\cdot}c_{sol}(90 - 44) = 1320 \text{ kJ/kg}$$

$$q_B = f{\cdot}c_{sol}(144 - 90) = 1540 \text{ kJ/kg}$$

  (ii) heat pump

$$COP = \frac{\bar{q}_B(abs) + q_B(102 \to 55) + \bar{q}A(cond)}{\bar{q}_B(des) + q_B(44 \to 90)}$$

$$= \frac{2880 + 1240 + 2490}{2970 + 1320} = 1.54$$

  (iii) heat transformer

$$COP = \frac{\bar{q}_B(abs) - q_B(102 \to 156)}{q_A(eva) + q_A(44 \to 90) + q_B(des) - q_B(144 \to 90)}$$

$$= \frac{2880 - 1430}{2400 + 190 + 2970 - 1540} = 0.36$$

4. Absorption Chillers

(c) 80°C

(d) for a reversible machine

$$COP(\text{Carnot}) = \frac{q_E}{q_G} = \frac{T_0(T_2 - T_1)}{T_2(T_1 - T_0)} = 1.18$$

(e) $1/r = 0.16$

(f) $f = (1 - x_p)/(x_r - x_p) = 12$ (subscripts p = poor and r = rich in respect to refrigerant content)

(g) $w_p = f \cdot v_s \cdot \Delta p = 12(55 - 9)10^2/1\ 7\ 10^3 = 0.03$ kJ/kg

(h) $q_E = r_E - c'(35 - 5) = 2.360$ kJ/kg

   $q_C = r_C + c''(70 - 35) = 2480$ kJ/kg

(i) $q_G = f\, c_{sol} (68 - 35) + r_C(1 + 1/r) + (f - 1)c_{sol}(80 - 68) = 3970$ kJ/kg

(j) $q_A = r_e(1 + 1/r) + (f - 1)c_{sol}(80 - 45) + (f - 1)c_{sol}(45 - 35) - c''(35 - 5) = 3930$ kJ/kg

   Energy balance is fulfilled within several percent.

(k) $q_E/q_G = 0.59$

(l) high pressure side

(m) $q_{HX} = (f - 1)c_{sol}(80 - 35) = 1090$ kJ/kg

(n) $COP = q_E/q_G - q_{HX} = 0.82$ $COP/COP_{Carnot} = 0.69$

5. Superposition

(b) $1 + 2; 2 + 3; 1 + 4$

(c) $\theta_1 + \theta_1 Q_2 = 1.08; \theta_2 + \theta_2 Q_3 = 1.36; \theta_1/(1 - \theta_1\theta_4) = 0.86$

(d) $2 + 3$

6. Superposition

Cycle a:

$$COP = \theta_1 + \theta_1^2 = 1.19$$

$$COP = \theta_1 + \theta_1^2 = 1.19$$

$$COP = \theta_1/(1 - \theta_1\theta_4) - 1.04$$

Cycle b:

$$COP = 2\theta_1 = 1.40$$

$$COP = \theta_1 + \theta_1^2 = 1.19$$

$$COP = \theta_1/(1 - \theta_2) = 1.32$$

Cycle c:

$$COP = 2\theta_1 = 1.40$$

$$COP = \theta_1 + \theta_1^2 = 1.19$$

$$COP = \theta_1/(1 - \theta_2) = 1.32$$

7. Entropy and Work

(a)	$Q = 0$	$W = mgh$	$\Delta S = 0$
(b)	$Q = mgh$	$W = 0$	$\Delta S = mgh/T$
(c)	$Q = 0$	$W = p\Delta V$	$\Delta S = 0$
(d)	$Q = 0$	$W = 0$	$\Delta S = 0$
(e)	$Q = 0$	$W = (p_h - p_c)\Delta V$	$\Delta S = 0$
(f)	$Q = C_V(T_2 - T_1)$	$W = 0$	$\Delta S = C_V \ln \dfrac{T_2}{T_1}$
(g)	$Q = 0$	$W = C_V(T_2 - T_1)$	$\Delta S = 0$
(h)	$Q = 0$	$W = C_P(T_2 - T_1)$	$\Delta S = 0$
(i)	$T_2^{(f)} = T_2^{(g)}; V_2^{(f)} > V_2^{(g)}$		

8. Entropy Analysis for a Vapor-Compression System

(b) COP = 2.50

(c)

$$\dot{W} = \dot{Q}\left(\frac{T_R - T_N}{T_N}\right) + T_R \Sigma \delta \dot{S}$$

(d) Evaporator: 73.4 mW/K; Condensor: 305.0 mW/K

Compressor: 382.7 mW/K; Desuperheating: 21.6 mW/K

Expansion valve: 188.0 mW/K; minimum work 114.1 W

(e) COP = 2.50

(f) COP = 4.42 for $T_K \rightarrow T_R$

Properties for R22 at saturation:

T[°C]	p[bar]	h'[kJ/kg]	h'[kJ/kg]	s'[kJ/kgK]	s"[kJ/kgK]
−15	2.96	182.16	399.16	0.9334	1.7741
20	9.081	224.34	411.14	1.0848	1.7221
40	15.27	249.22	415.19	1.1651	1.6952

Properties for superheated R22:

p[bar]	s[kJ/kgK]	T[°C]	h[kJ/kg]
9.081	1.7741	65.9	426.9
15.27	1.7741	65.9	441.1

9. Vapor-Compression Refrigeration Unit

(b) P_0: $h' = 0$ $h'' = 221$ kJ/kg

$s' = 0$ $s'' = 0.873$ kJ/kgK

P_1: $h' = 84$ kJ/kg $h' = 238$ kJ/kg

   $s' = 0.293$ kJ/kgK $s'' = 0.770$ kJ/kgK

(c) isentropic compression

   $s_{is} = s''(P_0)$; $T_{is} = T \exp[(s'(P_0) - s''(P_1))/c_p''] = 85.2°C$

   $h_{is} = 273$ kJ/kg

   $W_{is} = 52$ kJ/kg

   real compression

   $W_{real} = W_{is}/g_c = 74.3$ kJ/kg

   $h = 295.3$ kJ/kg

   $T = 107.3°C$

   $s = 0.933$ kJ/kg

(d)

$$\frac{m(vapor)}{m(total)} = \frac{c_p'(T_1 - T_0)}{h''(P_0) - h'(P_0)} = 0.38$$

(e) cooling: 137 kJ/kg

   heating: 211 kJ/kg

(f) cooling: 1.84

   heating: 2.84

(g) The value of the higher temperature depends on the point of view.

10. Vapor-Compression Refrigeration System

(b) case a: 0.14 m^3/sec

   case b: 0.125 m^3/sec

       0.045 m^3/sec

(c) case a: 1.67

   case b: 1.85

11. Comparison of Absorption and Vapor-Compression Refrigeration Units

(a) COP = $T_0/(T_1 - T_0) = 9.3$

(b) $1/T_2 = 1/T_1 - r_R/r_A (1/T_0 - 1/T_1)$; $T_2 = 68°C$

(c) COP = $r_R/r_A = 0.9$

(d) $COP_{real} \approx 2.6$ (compr.)

   $COP_{real} \approx 0.7$ (absorb, with $H_2O/LiBr$)

12. Gas Cycle Refrigeration System

(b)
$$\Pi_{min} = \left(\frac{T_1}{T_3 \eta_2} - \frac{1-\eta_2}{\eta_2} \right)^{\kappa/(1-\kappa)} = 1.94$$

(c)
$$COP = \frac{T_1 - T_3 \eta_2 \Pi^{1-\kappa/\kappa} - T_3 (1-\eta_2)}{\frac{T_1}{\eta_1} \left(\Pi^{(\kappa-1)/\kappa} - 1 \right) - T_3 \eta_2 \left(1 - \Pi^{(1-\kappa)/\kappa} \right)}$$

$$COP_{rev} = \frac{1}{\Pi^{\kappa-1/\kappa} - 1} = 5.46 \text{ for } \Pi = \Pi \text{ min}$$

$$\Pi_{opt} = 4.3 \qquad COP_{opt} = 0.48$$

(e)
$$\Pi = \left(\frac{T_5}{T_3 \eta} - \frac{1-\eta_2}{\eta_2} \right)^{\kappa/(1-\kappa)} = 1.68$$

(f)
$$COP = \frac{T_0 - T_5}{\frac{T_0}{\eta_1} \left(\Pi^{(\kappa-1/\kappa)} - 1 \right) - \left(T_3 - T_5 \right)} = 0.95$$

13. Rankine Cycle in Various Diagrams
Before sketching the cycles prepare qualitative diagrams!

14. Steam Power Plant

(a)
$$\overline{T} = \frac{\Delta Q}{\Delta S} = \frac{c'(T_e - T_0) + r + c''(T_2 - T_e)}{c' \ln \frac{T_e}{T_0} + \frac{r}{T_e} + c'' \ln \frac{T_2}{T_e}} = 259°C$$

with second superheat:
$$\overline{T} = \frac{\Delta Q + c''(T_4 - T_3)}{\Delta S + c'' \ln \frac{T_4}{T_3}}$$

(b)
$$Hint. \quad COP - 1 = \frac{T_0}{\overline{T}}$$

(c)
$$T_3(opt) = \overline{T}(T_3)$$
$$T_3 = 279°C; P(opt) = 28.9 \text{ bar}$$

(d)
$$0.45; \quad 0.43$$

15. Internal Combustion Engine Driven Heat Pump

(b) $COP_{HP} = 3.77$

(c) 1.86 vs 1.32

(d) 54% vs 36%

(e) yes: for diesel-engine plus HP

no: for power plant plus HP

16. Cogeneration of Heat and Power

(a) A: $\Delta E/P = 1/\eta_B\left(\eta_S\left(1-\eta_M\right)/\eta_M+\varepsilon\right)-1/\eta_M=2.13$

B: $\Delta E/P = 1/\eta+\eta_S/\eta_B\left(1-\eta_M\right)\eta_M-1/\eta_M=1.86$

(c) Compare the waste heat streams.

17. Power-Refrigeration Cogeneration

(b) $$\eta=\eta_2+\eta_1\left(1-\eta_2\right)$$

(c) Option A : $\dfrac{W}{P}=\dfrac{1}{\eta\eta_P}+\dfrac{1}{\eta\eta_P\cdot COP_r}\cdot\dfrac{R}{P}=2.77+0.6\dfrac{R}{P}$

Option B : $\dfrac{W}{P}=\dfrac{1}{\eta\eta_P}+\dfrac{\eta_1}{\eta\eta_H COP_a}$

$=2.77+0.49\dfrac{R}{P}$ (single effect)

$=2.77+0.57\dfrac{R}{P}$ (double effect)

$\left(\dfrac{R}{P}\right)_{max}=\dfrac{\eta_H}{\eta_P}\dfrac{1-\eta}{\eta-\eta_1}COP_a$

$=2.07$ (single effect)

$=6.44$ (double effect)

18. Analysis of Heat Exchanger Networks with the Pinch Method

(a) recuperated heat: 21.3 MW

heating load: 2.2 MW

cooling load: 5.7 MW

(b) recuperated heat: 18.0 MW

heating load: 4.6 MW

cooling load: 8.1 MW

Tables Representing Opportunities for Internal Heat Exchange

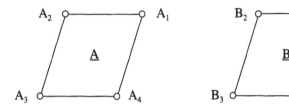

Configurations with two basic absorption cycles, A and B.

TYPE OF COUPLING A_1*B_1

Heat exchange on 1 temperature level in addition to the common exchange unit. If the common exchange unit is a separate entity to each cycle (Cascade), then the two exchange units are operating on the temperature level and are in heat exchange with each other and/or an external reservoir. This statement holds for all cycles presented here.

$B_1A_1 (+ -)$	$B_2A_2 (+ -)$	$B_3A_2 (+ +)$	$B_4A_2 (+ -)$
	$B_2A_3 (+ +)$	$B_3A_3 (+ -)$	$B_4A_3 (+ +)$
	$B_2A_4 (+ -)$	$B_3A_4 (+ +)$	$B_4A_4 (+ -)$

	$B_2A_2A_4 (+ - -)$	$B_3A_2A_4 (+ + +)$	$B_4A_2A_4 (+ - -)$
	$A_2B_2B_4 (+ - -)$	$A_3B_2B_4 (+ + +)$	$A_4B_2B_4 (+ - -)$

Heat exchange on 2 temperature levels.

$B_1A_1 (+ -)$ and $B_2 A_2 (+ -)$ $B_1A_1 (+ -)$ and $B_2 A_2 A_4 (+ - -)$
$B_1A_1 (+ -)$ and $B_2 A_4 (+ -)$ $B_1A_1 (+ -)$ and $A_2 B_2 B_4 (+ - -)$
$B_1A_1 (+ -)$ and $B_3 A_3 (+ -)$ $B_1A_1 (+ -)$ and $B_4 A_2 A_4 (+ - -)$
$B_1A_1 (+ -)$ and $B_4 A_2 (+ -)$ $B_1A_1 (+ -)$ and $A_4 B_2 B_4. (- + +)$
$B_1A_1 (+ -)$ and $B_4 A_4 (+ -)$

TYPE OF COUPLING A_1*B_2

Heat exchange on 1 temperature level (in addition to the common exchange unit).

$B_2 A_1 (+ +)$	$B_3 A_2 (+ +)$	$B_4 A_1 (+ +)$
	$B_3 A_3 (+ -)$	$B_4 A_2 (+ -)$
	$B_3 A_4 (+ +)$	$B_4 A_3 (+ +)$
		$B_4 A_4 (+ -)$

$B_2 A_1 B_4 (+ + +)$ $B_3 A_2 A_4 (+ + +)$ $B_4 A_2 A_4 (+ - -)$

Heat exchange on 2 temperature levels.

B_2A_1 (+ +) and B_3 A_2 (+ +) B_2 A_1 (+ +) and B_3 A_2 A_4 (+ + +)
B_2A_1 (+ +) and B_3 A_4 (+ +) B_3 A_2 (+ +) and B_4 A_1 B_2 (+ + +)
B_2A_1 (+ +) and B_4 A_3 (+ +) B_3 A_4 (+ +) and B_4 A_1 B_2 (+ + +)

B_3 A_3 (+ –) and B_4 A_2 (+ –) B_2 A_1 B_4 (+ + +) and B_3 A_2 A_4 (+ + +)
B_3 A_3 (+ –) and B_4 A_4 (+ –)

TYPE OF COUPLING A_1*B_3

A_1 B_3 (+ +)

TYPE OF COUPLING A_1*B_4

Heat exchange on 1 temperature level (in addition to the common exchange unit).

B_2 A_1 (+ +) B_3A_2 (+ +) B_4 A_1 (+ +)
B_2 A_2 (+ –) B_3A_3 (+ –)
B_2 A_3 (+ +) B_3A_4 (+ +)
B_2 A_4 (+ –)

$B_2B_4A_1$ (+ + +) $B_3A_2A_4$ (+ + +) $B_2A_2A_4$ (+ – –)

Heat exchange on 2 temperature levels.

B_4 A_1 (+ +) and B_2 A_3 (+ +) B_2 A_1 (+ +) and B_3 A_2 (+ +)
B_4 A_1 (+ +) and B_3 A_2 (+ +) B_2 A_1 (+ +) and B_3 A_4 (+ +)
B_4 A_1 (+ +) and B_3 A_4 (+ +) B_2 A_2 (+ –) and B_3 A_3 (+ –)
                                                B_2 A_4 (+ –) and B_3 A_3 (+ –)

B_4 A_1 (+ +) and B_3 A_2 A_4 (+ + +)
B_2 A_1 (+ +) and B_3 A_2 A_4 (+ + +)

B_2 B_4 A_1 (+ + +) and B_3 A_2 A_4 (+ + +)

TYPE OF COUPLING A_2*B_2

Heat exchange on 1 temperature level (in addition to the common exchange unit).

B_1 A_1 (+ –) B_2 A_2 (+ –) B_3 A_3 (+ –) B_4 A_1 (+ +)
B_1 A_4 (+ +) B_2 A_4 (+ –) B_3 A_4 (+ +) B_4 A_2 (+ –)
                                                                        B_4 A_3 (+ +)
                                                                        B_4 A_4 (+ –)

B_2 A_2 A_4 (+ –) B_2 A_2 B_4 (+ –) B_2 A_2 B_4 A_4 (– – + –)

Heat exchange on 2 temperature levels.

B_2 A_2 (– –) and B_1 A_1 (+ –) B_1 A_1 (+ –) and B_2 A_2 B_4 (+ – +)
B_2 A_2 (+ –) and B_2 A_4 (+ –) B_3 A_3 (+ –) and B_2 A_2 B_4 (+ – +)
B_2 A_2 (+ –) and B_3 A_3 (+ –) B_1 A_1 (+ –) and B_2 A_2 A_4 (+ – –)

$B_2 A_2 (+ -)$ and $B_4 A_4 (+ -)$ $B_3 A_3 (+ -)$ and $B_2 A_2 A_4 (+ - -)$

$B_1 A_4 (+ +)$ and $B_4 A_3 (+ +)$
$B_3 A_4 (+ +)$ and $B_4 A_1 (+ +)$

TYPE OF COUPLING A_2*B_3

Heat exchange on 1 temperature level (in addition to the common exchange unit).

$B_1 A_1 (+ -)$ $B_2 A_1 (+ +)$ $B_3 A_2 (+ +)$ $B_4 A_1 (+ +)$
$B_1 A_4 (+ +)$ $B_2 A_4 (+ -)$ $B_3 A_4 (+ +)$ $B_4 A_4 (+ -)$

$B_3 A_2 A_4 (+ + +)$ $B_2 A_1 B_4 (+ + +)$ $B_2 A_4 B_4 (+ - +)$

Heat exchange on 2 temperature levels.

$B_3 A_2 (+ +)$ and $B_1 A_4 (+ +)$
$B_3 A_2 (+ +)$ and $B_2 A_1 (+ +)$
$B_3 A_2 (+ +)$ and $B_4 A_1 (+ +)$

$B_1 A_1 (+ -)$ and $B_2 A_4 (+ -)$
$B_1 A_1 (+ -)$ and $B_4 A_4 (+ -)$

$B_2 A_1 (+ +)$ and $B_3 A_2 A_4 (+ + +)$
$B_4 A_1 (+ +)$ and $B_3 A_2 A_4 (+ + +)$
$B_3 A_2 (+ +)$ and $B_2 A_1 B_4 (+ + +)$
$B_3 A_4 (+ +)$ and $B_2 A_1 B_4 (+ + +)$
$B_1 A_1 (+ -)$ and $B_2 A_4 B_4 (+ - +)$

$B_2 A_1 B_4 (+ + +)$ and $B_3 A_2 A_4 (+ + +)$

TYPE OF COUPLING A_2*B_4

Heat exchange on 1 temperature level (in addition to the common exchange unit).

$B_1 A_1 (+ -)$ $B_2 A_1 (+ +)$ $B_3 A_3 (+ -)$ $B_4 A_2 (+ +)$
$B_1 A_4 (+ +)$ $B_2 A_2 (+ -)$ $B_3 A_4 (+ +)$ $B_4 A_4 (+ -)$
                    $B_2 A_3 (+ +)$
                    $B_2 A_4 (+ -)$

$B_4 A_2 B_2 (+ - +)$ $B_4 A_2 A_4 (+ - -)$ $B_4 A_2 B_2 A_4 (- - - -)$

Heat exchange on 2 temperature levels.

$B_4 A_2 (+ -)$ and $B_3 A_3 (+ -)$ $B_1 A_1 (+ -)$ and $B_2 A_2 B_4$
$B_4 A_2 (+ -)$ and $B_2 A_4 (+ -)$ $B_1 A_1 (+ -)$ and $B_4 A_2 A_4$
$B_4 A_2 (+ -)$ and $B_1 A_1 (+ -)$ $B_3 A_3 (+ -)$ and $B_2 A_2 B_4$
                                         $B_3 A_3 (+)$ and $B_4 A_2 A_4$

$B_1 A_4 (+ +)$ and $B_2 A_3 (+ +)$
$B_2 A_1 (+ +)$ and $B_3 A_4 (+ +)$

TYPE OF COUPLING A_3*B_3

Heat exchange on 1 temperature level (in addition to the common exchange unit).

$B_1 A_1 (+ -)$	$B_2 A_1 (+ +)$	$B_3 A_3 (+ -)$	$B_4 A_1 (+ +)$
$B_1 A_2 (+ +)$	$B_2 A_2 (+ -)$		$B_4 A_2 (+ -)$
$B_1 A_4 (+ +)$	$B_2 A_4 (+ -)$		$B_4 A_4 (+ -)$

$B_1 A_2 A_4 (+ + +)$	$B_2 A_2 A_4 (+ - -)$	$B_4 A_2 A_4 (+ - -)$
$A_1 B_2 B_4 (+ + +)$	$A_2 B_2 B_4 (+ - -)$	$A_4 B_2 B_4 (+ - -)$

Heat exchange on 2 temperature levels.

$B_3 A_3 (+ -)$ and $B_1 A_1 (+ -)$
$B_3 A_3 (+ -)$ and $B_2 A_2 (+ -)$
$B_3 A_3 (+ -)$ and $B_2 A_4 (+ -)$
$B_3 A_3 (+ -)$ and $B_4 A_2 (+ -)$
$B_3 A_3 (+ -)$ and $B_4 A_4 (+ -)$

$B_3 A_3 (+ -)$ and $B_2 A_2 A_4 (+ - -)$
$B_3 A_3 (+ -)$ and $A_2 B_2 B_4 (- + +)$
$B_3 A_3 (+ -)$ and $B_4 A_2 A_4 (+ - -)$
$B_3 A_3 (+ -)$ and $A_4 B_2 B_4 (- + +)$

TYPE OF COUPLING A_3*B_4

Heat exchange on 1 temperature level (in addition to the common exchange unit).

$B_1 A_1 (+ -)$	$B_2 A_1 (+ +)$	$B_4 A_3 (+ +)$
$B_1 A_2 (+ +)$	$B_2 A_2 (+ -)$	
$B_1 A_4 (+ +)$	$B_2 A_3 (+ +)$	
	$B_2 A_4 (+ -)$	

$B_1 A_2 A_4 (+ + +)$	$B_2 A_2 A_4 (+ - -)$	$B_2 B_4 A_3 (+ + +)$

Heat exchange on 2 temperature levels.

$B_4 A_3 (+ +)$ and $B_1 A_2 (+ +)$
$B_4 A_3 (+ +)$ and $B_1 A_4 (+ +)$
$B_4 A_3 (+ +)$ and $B_2 A_1 (+ +)$

$B_1 A_1 (+ -)$ and $B_2 A_2 (+ -)$
$B_1 A_1 (+ -)$ and $B_2 A_4 (+ -)$

$B_4 A_3 (+ +)$ and $B_1 A_2 A_4 (+ + +)$
$B_1 A_2 (+ +)$ and $B_2 A_3 B_4 (+ + +)$
$B_1 A_4 (+ +)$ and $B_2 A_3 B_4 (+ + +)$

$B_1 A_2 A_4 (+ + +)$ and $B_2 B_4 A_3 (+ + +)$

TYPE OF COUPLING $A_4{}^*B_4$

Heat exchange on 1 temperature level (in addition to the common exchange unit).

$B_1 A_1 (+ -)$	$B_2 A_1 (+ +)$	$B_3 A_2 (+ +)$	$B_4 A_2 (+ -)$
$B_1 A_2 (+ +)$	$B_2 A_2 (+ -)$	$B_3 A_3 (+ -)$	$B_4 A_4 (+ -)$
	$B_2 A_3 (+ +)$		
	$B_2 A_4 (+ -)$		

$B_2 B_4 A_4 (+ + -)$ $B_4 A_2 A_4 (+ - -)$ $B_4 A_4 B_2 A_2 (+ - + -)$

Heat exchange on 2 temperature levels.

$B_4 A_4 (+ -)$ and $B_1 A_1 (+ -)$ $B_1 A_1 (+ -)$ and $B_2 B_4 A_4 (+ + -)$
$B_4 A_4 (+ -)$ and $B_2 A_2 (+ -)$ $B_1 A_1 (+ -)$ and $B_4 A_2 A_4 (+ - -)$
$B_4 A_4 (+ -)$ and $B_3 A_3 (+ -)$ $B_3 A_3 (+ -)$ and $B_2 B_4 A_4 (+ - -)$
                                          $B_3 A_3 (+ -)$ and $B_4 A_2 A_4 (+ - -)$

$B_1 A_2 (+ +)$ and $B_2 A_3 (+ +)$
$B_2 A_1 (+ +)$ and $B_3 A_2 (+ +)$

TABLES REPRESENTING OPPORTUNITIES FOR INTERNAL HEAT EXCHANGE

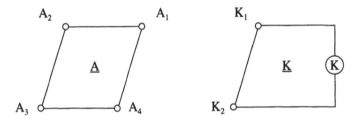

Configurations with one basic absorbtion cycle and one basic compression cycle, A and K.

TYPE OF COUPLING $A_1{}^*K_1$

Heat exchange on 1 temperature level.

$K_1 A_1 (+ +)$ $K_2 A_2 (+ +)$
                      $K_2 A_3 (+ -)$
                      $K_2 A_4 (+ +)$

$K_2 A_2 A_4 (+ + +)$

Heat exchange on 2 temperature levels.

$K_1 A_1 (+ +)$ and $K_2 A_2 (+ +)$
$K_1 A_1 (+ +)$ and $K_2 A_4 (+ +)$
$K_1 A_2 (+ +)$ and $K_2 A_2 A_4 (+ + +)$

TYPE OF COUPLING A_1*K_2

$K_2 \, A_1 \, (+ \, -)$

TYPE OF COUPLING A_2*K_1

Heat exchange on 1 temperature level.

$K_1 \, A_2 \, (+ \, -)$ $K_2 \, A_3 \, (+ \, -)$
$K_1 \, A_4 \, (+ \, -)$ $K_2 \, A_4 \, (+ \, +)$

$K_1 \, A_2 \, A_4 \, (+ \, - \, -)$

Heat exchange on 2 temperature levels.

$K_1 \, A_2 \, (+ \, -)$ and $K_2 \, A_3 \, (+ \, -)$
$K_2 \, A_3 \, (+ \, -)$ and $K_1 \, A_2 \, A_4 \, (+ \, - \, -)$

TYPE OF COUPLING A_2*K_2

Heat exchange on 1 temperature level.

$K_1 \, A_1 \, (+ \, +)$ $K_2 \, A_2 \, (+ \, +)$
$K_1 \, A_4 \, (+ \, -)$ $K_2 \, A_4 \, (+ \, +)$

$K_2 \, A_2 \, A_4 \, (+ \, + \, +)$

Heat exchange on 2 temperature levels.

$K_2 \, A_2 \, (+ \, +)$ and $K_1 \, A_1 \, (+ \, +)$

$K_1 \, A_1 \, (+ \, +)$ and $K_2 \, A_2 \, A_4 \, (+ \, + \, +)$

TYPE OF COUPLING A_3*K_1

$A_3 \, K_1 \, (+ \, +)$

TYPE OF COUPLING A_3*K_2

Heat exchange on 1 temperature level.

$K_1 \, A_2 \, (+ \, -)$ $K_2 \, A_3 \, (+ \, -)$
$K_1 \, A_4 \, (+ \, -)$
$K_1 A_1 \, (+ \, +)$

$K_1 \, A_2 \, A_4 \, (+ \, - \, -)$

Heat exchange on 2 temperature levels.

$K_2 \, A_3 \, (+ \, -)$ and $K_1 \, A_2 \, (+ \, -)$
$K_2 \, A_3 \, (+ \, -)$ and $K_1 \, A_4 \, (+ \, -)$

$K_2 \, A_3 \, (+ \, -)$ and $K_1 \, A_2 \, A_4 \, (+ \, - \, -)$

TYPE OF COUPLING $A_4 * K_1$

Heat exchange on 1 temperature level.

$K_1 A_2 (+ -)$ $K_2 A_2 (+ +)$
$K_1 A_4 (+ -)$ $K_2 A_3 (+ -)$

$K_1 A_2 A_4 (+ - -)$

Heat exchange on 2 temperature levels.

$K_1 A_4 (+ -)$ and $K_2 A_3 (+ -)$
$K_2 A_3 (+ -)$ and $K_1 A_2 A_4 (+ - -)$

TYPE OF COUPLING $A_4 * K_2$

Heat exchange on 1 temperature level.

$K_1 A_2 (+ -)$ $K_2 A_2 (+ +)$
$K_1 A_1 (+ +)$ $K_2 A_4 (+ +)$

$K_2 A_2 A_4 (+ + +)$

Heat exchange on 2 temperature levels.

$K_2 A_4 (+ +)$ and $K_1 A_1 (+ +)$
$K_1 A_1 (+ +)$ and $K_2 A_2 A_4 (+ + +)$

Part Load Characteristics and Optimization

To describe part load performance of an absorption heat pump, a rather simple relationship can be derived. This relationship describes the dependence of the capacity from external temperatures, i.e. the temperature of the heat sink and source fluids. In a ln(P), (-1)/T diagram, the vapor pressure lines for pure water and a LiBr-water mixture are essentially straight lines (Figure 1). As described by Clapeyron's equation, the slope of the vapor pressure line for water is proportional to the latent heat of water, r. For the salt solution, the slope is proportional to the latent heat of the water plus the heat of mixing (r+l). The amount of l depends on the salt and the concentration. In the case of LiBr-water, the amount of l is small compared to r. Thus the vapor pressure lines and their slope determine a relationship between T_E and T_C and also between T_A and T_G as shown in the next equation. The subscripts have the following meaning: E = evaporator, C = condenser, A = absorber and G = generator.

$$r\left(\frac{1}{T_E} - \frac{1}{T_C}\right) = (r+l)\left(\frac{1}{T_A} - \frac{1}{T_G}\right) \tag{1}$$

Rearranging the equation, one obtains

$$R\left(T_A - T_G\right) - \left(T_E - T_C\right) = 0 \tag{2}$$

with

$$R = \frac{r+l}{r}\frac{T_E T_C}{T_A T_G} \tag{3}$$

The temperatures T_i are internal temperatures in K, i.e. the temperatures of the working fluid. For a LiBr water system (r+l)/r is about 1.15 and R about 0.9. Now, the following temperature difference ΔT_t is introduced for the external temperatures.

$$R = \left(t_G - t_A\right)\left(t_C - t_E\right) = \Delta T_i \tag{4}$$

The t_i are mean values of the gliding temperatures for each of the heat transfer streams. The difference $(t_G - t_A)$ represents the difference between generator and absorber, the driving temperature difference. This temperature difference is a measure for the driving potential. On the other hand, there is a work consuming refrigeration process pumping heat q_0 from t_E to t_C. The according temperature difference, $(t_C - t_E)$, is called the temperature lift. By adding Equation 2 to Equation 1, an equation is obtained that contains the temperature differences between internal and external temperatures:

$$\left(T_C - t_C\right) + \left(t_E - T_E\right) + R\left(T_A - t_A\right) + R\left(t_G - T_G\right) = \Delta T_t \tag{5}$$

The internal and external temperatures are coupled through the heat exchangers. Now a relationship between the logarithmic mean temperature difference and the temperature

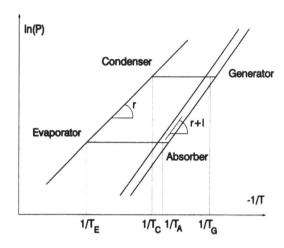

Figure 1.

differences (T_i-t_i) has to be found. In case of large differences, the approximation that ΔT_{log} is approximately equal to the absolute value of ABS(T_i-t_i) is valid. Thus it is assumed that the heat transfer relationships are written as

$$\dot{Q}_E = k_E A_E \left(t_E - T_E\right) a_E \tag{6}$$

$$\dot{Q}_C = k_C A_C \left(t_C - T_C\right) a_C \tag{7}$$

$$\dot{Q}_A = k_A A_A \left(t_A - T_A\right) a_A \tag{8}$$

$$\dot{Q}_G = k_G A_G \left(t_G - T_G\right) a_G \tag{9}$$

The Q_i are the amounts of heat exchanged at the respective overall heat exchanger, k_i the heat transfer coefficients, A_i the heat exchanger areas and the a_i correction factors for the fact that the temperature differences are approximations. In the following, these factors are assumed to be equal to 1.0. Combining Equations 5 through 9 and using $\eta = Q_E/Q_G$ and the assumption $Q_E = Q_C$ one can find for the cooling capacity

$$\dot{Q}_E = \cfrac{k_E A_E \Delta T_t}{1 + \cfrac{k_E A_E}{k_C A_C} + \cfrac{R}{\eta}\left(\cfrac{k_E A_E}{k_A A_A} + \cfrac{k_E A_E}{k_G A_G}\right)} \tag{10}$$

Accordingly all other heat quantities can be derived. The total area of all four heat exchangers is

$$A = A_E + A_C + A_A + A_G \tag{11}$$

The total investment cost for all four heat exchangers can be written as

$$A = m_E A_E + m_C A_C + m_A A_A + m_G A_G \tag{12}$$

with the m_i representing the specific cost per unit area.

Using Equation 10, one can estimate how the capacity of a system changes with small changes in the external temperatures. The important term in Equation 10 is ΔT_t as long as the changes in η are small. The latter was confirmed in experiments to hold very well. Assuming that a single stage LiBr water machine is designed for the following temperatures: $t_G = 90°C$, $t_A = t_C = 35°C$ and $t_E = 8°C$, then ΔT_t amounts to 22.5K (with R = 0.9). In case the heat rejection temperature is lowered from 35°C to 32°C, then ΔT_t increases to 28.2K and thus the capacity by 25%. If the temperature of the heat source is reduced by 3K, then ΔT_t = 19.8K and the capacity decreases by 12%. In case the capacity is required to remain constant, then the evaporator temperature has to increase by 3K or the temperature of the cooling water has to be reduced by 1.4K or the areas have to be increased by 12%. These values are independent of the a_i as long as those do not vary much.

Equation 10 can also be used to calculate the optimum allocation of surface area to the heat exchangers for a given total area or for a given total cost M. It is assumed that the overall heat transfer coefficients have the following values: $k_C = k_E = 4000W/m^2K$ and $k_A = k_G = 1000W/m^2K$. It is further assumed that the efficiency is $\eta = 0.7$ and the $\Delta T_t = 22.5K$. Lastly it is assumed that the total area is distributed evenly for all heat exchangers.

$$\frac{A}{4} = A_E = A_C = A_A = A_G \tag{13}$$

The denominator in Equation 10 amounts to 12.3. For A = 40m² the evaporator capacity is 73.2kW. The condenser capacity has the same value and the absorber and generator capacities are increased by $1/\eta$. They amount to 104.5kW. The temperature difference t_E-T_E amounts to 1.83K according to Equations 6 and 10. All other temperature differences can be calculated according to Equations 7 through 9.

t_E-T_E	=	1.83K	Q_E = 73.2kW
T_C-t_C	=	1.83K	Q_C = 73.2kW
T_A-t_A	=	10.49K	Q_A = 104.5kW
t_G-T_G	=	10.49K	Q_G = 104.5kW

When the values for the temperature differences are used in Equation 5, one obtains for ΔT_t = 22.5K as it should be the case.

The maximum evaporator capacity can be found by building the derivative of Equation 10 with respect to A_i with the boundary condition A = const. = ΣA_i. The result is:

$$A_i = \frac{A}{N} \sqrt{\frac{\alpha_i}{k_i}} \text{ with } i = A, C, E, G \tag{14}$$

and with

$$N = \Sigma_i \sqrt{\frac{\alpha_i}{k_i}} \text{ and } \alpha_E = \alpha_C = 1 \quad \alpha_G = \alpha_A = \frac{R}{\eta} \tag{15}$$

N amounts to 3.266 and for optimum area allocation $A_E = A_C = 0.153A = 6.1m^2$ and $A_G = A_A = 0.347A = 13.9m^2$. The denominator in Equation 10 is now equal to 6.54. Thus one obtains for the capacities and temperature differences the following values:

t_E-T_E	= 3.44K	Q_E = 84.0kW
T_C-t_C	= 3.44K	Q_C = 84.0kW
T_A-t_A	= 8.65K	Q_A = 120.0kW
t_G-T_G	= 8.65K	Q_G = 120.0kW

The capacity of the machine increases by 15% without increasing the heat exchange area. The area was reallocated in favor of the generator and absorber.

If the optimization were conducted at a fixed total cost and were the factors a_i in equation not equal to 1.0, the following values were obtained for the factors α_i:

$$\alpha_i = \frac{1}{a_i m_i} \quad \text{with } i = E, C \tag{16}$$

and

$$\alpha_i = \frac{R}{\eta a_i m_i} \quad \text{with } i = A, G \tag{17}$$

Depending on the specific costs, a different optimum surface area allocation will be found.

Similar relationships (Equations 1 through 4) can be found for multistage systems. For the double effect system, the ΔT_t can be written as in Equation 4 with the following modification

$$R(\text{double effect}) = \frac{\left(\dfrac{r+1}{r}\right)^2 \dfrac{T_E T_C}{T_A T_G}}{1 + \dfrac{r+1}{r}} \tag{18}$$

This factor amounts to about half of the value for a single stage system. The sensitivity to the temperatures of the heat transfer media is changed. Increasing the temperature of the cooling water by 1K requires an increase of 3K for the generator.

In conclusion, the following aspects are noted:

1. Equation 1 can also be derived using an entropy balance. This has the advantage that the average values $1/T_i$ ($i = A,C,E,G$) in Equation 1 are calculated exactly. This derivation shows the justification for the use of temperature values of the middle of the temperature glide.

2. It is more difficult to develop a relationship between the entropic temperature averages and the driving temperature difference for the heat exchangers i.e the determination of the factors a_i in Equations 6 through 9. No general statement can be made because the value of the a_i depends not only on design details but also on what temperatures are being selected for the definition and measurement of the heat transfer coefficients. For a counter flow heat exchanger with respective temperature glides of d and d', the a_i can be calculated as a function of the driving temperature difference Δ (Logarithmic mean temperature difference) according to the following equation (see Figure 2).

$$a = \frac{2\Delta}{d-d'} \frac{1 - \exp\left(-\dfrac{d-d'}{\Delta}\right)}{1 + \exp\left(-\dfrac{d-d'}{\Delta}\right)} \tag{19}$$

This equation can be derived by inserting the entropic temperature average as the average value of the individual temperature glides into the logarithmic mean temperature difference for a counter flow heat exchanger. It can be seen that the factor a_i approaches 1.0 very quickly when the increasing driving temperature difference becomes larger than half of the difference of the temperature glide. In case an a_i is obtained that is significantly smaller than 1.0, than the estimates given above can be improved with an iterative scheme.

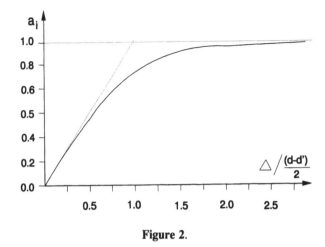

Figure 2.

REFERENCES

1. Takada, S. (1982) *Absorption Refrigeration Machine* (Textbook in Japanese), Editor Japanese Association of Refrigeration, Tokyo.

2. Furukawa, T. et al. (1983) Study on Characteristic Temperatures of Absorption Heat Pumps. Proc. 20th Japanese Heat Transfer Conference, June 1983, pp. 508-510 (in Japanese).

3. Riesch P. (1986) Aufbau und Betrieb eines Absorptionswaermetransformators, Master Thesis, Technische Universitaet Muenchen, Germany (in German).

INDEX

A

Absorbents, 31–36, 46
  solid, 89, 241, 244–246
Absorber-desorber heat exchange, 104–111
  COP, 110
  GAX cycles, 112, 170, 250
Absorption-compression systems, 123; See also
      Compression/expansion systems;
      Vapor-compression heat pumps
Absorption cycles, 37, 46–50, 85–90
  graphic representation of, 128, 131
Absorption-desorption cycles, 37, 50–52
  COP enhancement, 104–111
  Rankine Sorption Cycle, 37, 50–52, 86–87, 190
  solid-vapor systems, 241, 247–251
  weak and strong absorbent heat exchange, 111
Absorption heat pumps, 83
  COP, 90–98, 112, 117–118, 187, 199–201
    effectiveness, 113
    enhancement of, 98–111
      desorption-resorption heat exchange,
          104–111
      heat of rectification, 98–103
      internal heat exchangers, 98
      solution recirculation, 103–104
      three-pass heat exchanger, 98
    First and Second Laws for, 113–117
  enthalpy concentration diagram, 29
  exchange unit pairs, 83–85
  First and Second Laws for, 113–117
  heat and work flows, 83–84
  multistage, See Multistage absorption systems
  open system, 256
  part load performance, 283–286
  performance evaluation
    with desorber-absorber heat exchange, 104–111
    with heat rectification, 99–103
    lithium bromide/water system, 94–98
  quality (effectiveness), 113, 115–118
  Rankine/Rankine Heat Pumps, 231
  refrigeration capacity, 116–117
  sign combinations, 7
  single-stage system limitations, 111
  solid absorbents, 241
  two-stage, See Two-stage absorption systems
  vapor pressure curves, 89
  working fluids for, 31–36
  zeolite heat storage device, 244–246
Absorption heat pump transformer
  COP, 196–197, 202
  First and Second Law analysis, 16–17
  operating range, 142
  two-stage system, 150, 165
Absorption heat transformers
  COP, 192, 201–202
  efficiency, 119
  entropy production, 119

evaluation, 192–194
heat and work sign combinations, 7
high temperature-lift units, 163–165
multistage, See Multistage absorption systems
open system, 256
solid absorbents, 241
two-stage, See Two-stage absorption systems
Additives, 32
Air-conditioning systems
  cascade configurations, 133
  double-effect absorption systems, 167
  open systems, 255
  zeolite heat pump, 245–246
Air flow rate, 255
Ammonia, COP parameters, 69
Ammonia/water
  corrosion inhibition, 31
  desorber-absorber heat exchange, 104
  double-effect cycle COP, 199
  enthalpy-concentration diagram, 21
  heat of rectification, 98–103
  heat pumping pressure levels, 32
  heat pump temperature lift, 193
  high-efficiency two-stage system, 158, 161
  pressure-temperature diagram, 25
  solution fields, 34
  two-stage system configuration, 148
  vapor-compression heat pump with solution
      circuit, 55–61
  vapor pressure curves, 46, 89
Ammonia/water/lithium bromide, solution fields, 34
Asymmetry factors, 191–194
Availability
  absorption system operations, 83
  solid absorbent storage system, 243
  transformation with paired exchange units, 62
Azeotropic mixtures
  pressure-temperature diagram, 24
  research challenges, 36
  temperature-concentration diagrams, 22–23

B

Basic cycles, 126
  absorption/compression/expansion system
      configurations, 210, 231
  graphic representation, 128
  multistage cycle decomposition, 184
Binary mixtures, See Working fluid mixtures
Boiling point
  elevation, 83
  enthalpy concentration diagram, 29
  temperature-concentration diagram, 19
  working fluid mixtures, 23, 46
Booster compression, 123
Branched GAX cycle, 170, 250
Building blocks, 124–125
Bypasses, 127, 168

C

Carnot efficiency, 65, 67
Carnot factors, 7
Cascade configurations, 126, 133
  classification, 137
  design rules, 127
  high-efficiency two-stage system, 158
  multistage absorption configurations, 187
    compression/expansion configuration, 238–239
    heat storage device, 247
    superposition rules, 194–195
Clapeyron's equation, 92, 283
Clausius-Clapeyron equation, 71
Coefficient of performance (COP)
  absorption heat pump, 90–98, 112, 117–118, 199–202
    enhancement, 98–111
    with internal combustion engine, 112
    with limited data, 113
    with rectification, 103
    with temperature lift, 201–202
  desorber-absorber heat exchange, 108
  double-effect cycles, 199, 202
  energy saving factor (e), 198–202
  enhancement methods
    desorption-absorption heat exchange, 104–111
    heat of rectification, 98–103
    internal heat exchangers, 98
    solution recirculation, 103–104
    superheat use, 225–227
    three-pass heat exchanger, 98
  First Law efficiency, 63
  gliding temperature intervals and, 48
  heat of solution and, 93
  heat transformer, 118–121
  multistage absorption configurations, 187, 191, 194
    double-effect chiller, 195–196
    heat pump, 199–201
    heat pump transformer, 196–197, 202
    heat transformer, 201–202
    refrigeration, 198–199
    resorption cycle, 198
  Second Law-based determinations, 62–63
    analytic equations, 70–75
    exergy comparison, 63, 75–77
    heat pumping, 68–69
    refrigeration, 63–68
    results for selected fluids, 69–70
  single-stage heat transformer, 192
  single-stage refrigerator, 190–191
  solid-vapor systems, 250
  thermodynamic diagrams, 19
  two-stage absorption systems, 147–148, 154
    compression/expansion system, 218, 225
  two temperature level heat conversion, 5
  vapor-compression heat pump, 45, 49–50, 111
    with internal combustion engine, 112
    with solution circuit, 52, 61, 217
  zeolite heat pump, 246

Cogeneration systems, 133
Complete cycle, 126
Composition variable, 19
Compression/expansion systems, See also Vapor-compression heat pumps
  configuration decomposition, 126
  graphic representation, 128
  multistage absorption-coupled configurations, See Multistage absorption/compression/expansion systems
  sign combinations, 210–213
  solid-vapor systems with, 251–253
  two-stage absorption-coupled configurations, See Two-stage absorption/compression/expansion systems
Compression ratio, 49, 61
Compressor quality factor, 66
Compressor work, 44–45, 48, 216
Condensation, 37, 46–50; See also Absorption
Condenser inlet, vapor quality at, 37
Cooling capacity, 116–117, 165–167, 225
COP, See Coefficient of performance
Corrosion inhibition, 31
Crystallization, 32, 34
Cycle definitions, 126

D

Decomposition, 126, 194–195
Dehumidification systems, 255
Design rules, 123
  building block representation, 124–125
  classification system, 123
  graphic representation
    irreversible/reversible graphs, 128–131
    topological equivalence, 126, 131–132
  heat conversion system configuration, 126–128
Desorber-absorber heat exchange, See Absorber-desorber heat exchange
Desorption, 37, 46–50
Dew point, 19, 29
Diesel engine, 112
Distillation equipment, enthalpy concentration diagram, 29
Double-effect systems, 158–159, 166–167, 180
  compressor operation, 237
  COP, 195–196, 199, 202
Double-lift cycles, 201
  double-effect system, 166
  heat transformer, 165
Driving heat, See Generator

E

Effectiveness, absorption heat pump, 115–118
Efficiency, See also Coefficient of performance; Heat ratio
  absorber/working fluid systems, 35
  compression heat pumping, 38, 68–69, 112
  heat engine, 6
  heat transformer, 119

double-lift unit, 165
 single-stage units, 164–165, 192–194
internal heat exchange and, 45, 127, 155
refrigeration, 114, 188–191
steam power plant, 42–43
temperature lift and, 163, 220
three temperature level heat pumps/transformers,
 8–9
two-stage absorption systems, 137, 140, 158–161
working fluid properties and, 19, 30
Electric power storage, 245
Energy-saving factor (e), 198–202
Enthalpy
 desorber-absorber heat exchange, 106–108
 entropy relationships, 77–78
 lithium bromide/water absorption system, 95–97
 two-stage absorption systems, 145, 153
Enthalpy-concentration diagrams, 19–21, 90–91, 145
Entropic temperature averages, 64, 66, 286
Entropy production
 absorption heat pump, 114–115
 absorption heat transformer, 119
 conventional exchanger, 3
 enthalpy relationships, 77–78
 exergy balances, 77
 Second Law expression, 2
 solution circuit, 116
 three-temperature level heat transformer, 9
 two temperature level heat engine, 6
Environmental considerations, 30
Evaporation, 37; See also Desorption
 Rankine Cycles, 46–50
 temperature-concentration diagram, 19
Evaporator capacity maximization, 285–286
Evaporator outlet, vapor quality at, 37
Exchange units, 123; See also Heat exchanger
 performance
 availability transformation, 63–68
 configuration decomposition, 126
 connections, 123
  design rules, 126–127
  paired systems, 83–85
  topological equivalence, 131
  two-stage systems, 168–170
 cycle definitions, 126
 pressure drop assumptions, 124
Exergy analysis, 63, 75–77

F

Feed water preheating, 46
First Law analysis, 1, 63
 heat pump COP evaluation, 113–117
 heat transformer COP evaluation, 118–121
 sign combinations
  many temperature levels, 12–17
  three temperature levels, 6–12
  two temperature levels, 2–6
Flow rate
 design rules, 126–128
 lithium bromide/water absorption system, 94–95

low temperatures, 236
open systems, 255
vapor-compression heat pump with solution
 circuit, 56–58
Four-stage absorption configurations, 186–187

G

GAX cycles, 112, 170, 250
Generator, 85
 desorber-absorber heat exchange, 104
 solution recirculation, 103–104
 two-stage absorption systems, 140
Generator-absorber heat exchange (GAX), 112, 170,
 250
Gliding temperature, See Temperature glide
Glycol systems, 34–35
Graphic representation, See also Thermodynamic
 diagrams
 irreversible/reversible graphs, 128–131
 topological equivalence, 126, 131–132
Graphite, 246
Group theory, 123

H

Halogenated hydrocarbons
 COP parameters, 74
 heat capacity/latent heat ratio, 30
 vapor-compression systems, 31
Heat, sign conventions, 1
Heat capacity/latent heat ratio, 30
Heat conversion systems, defined, 1
Heat engine, 1
 evaluation, 38, 42–44
 heat and work sign combinations, 6
 Rankine Cycle, 37
Heat exchange, graphic representation of, 130
Heat exchanger performance, See also Exchange
 units
 lithium bromide/water absorption system, 95
 surface area allocation, 285–286
 two-stage absorption systems, 146–147, 152
 vapor-compression heat pump with solution
  circuit, 59–60
Heating capacity, two-stage absorption systems,
 165–167
Heat of absorption, 158
Heat of mixing, 93
Heat of rectification, 98–103, 144–147
Heat of vaporization, 116
Heat pump, 1; See also Absorption heat pumps;
 Vapor-compression heat pumps
Heat pump transformer, See Absorption heat pump
 transformer
Heat ratio, See also Coefficient of performance;
 Efficiency
 multistage cycles, 194
 three temperature-level systems, 8–9
 two-stage absorption systems, 139–142,
 158–161

Heat storage, 241; See also Solid-vapor systems
   compression/expansion machines, 251–253
   multistage systems, 26–250
   operating principles, 241–243
   sensible heat and, 244
   work and availability, 243
   zeolite device, 244–246
Heat transformation theory, 1
   work and heat sign combinations, 2–17
Heat transformer, See Absorption heat transformer
Hydrogen gas production, 62
Hydrogen refrigerant, 89

I

Internal combustion engine-driven systems, 11, 112,
   220, 237
Internal heat exchange
   configuration tables, 275–281
   COP enhancement methods, 98
   design rules, 127–128
   desorber-absorber heat exchange, 104–111
   graphic representation of, 130
   inter-configurational class switching, 170
   solid-vapor system, 247, 250–251
   solution recirculation, 103–104
   two-stage absorption systems, 137, 140, 154–157
     compression/expansion unit, 222–225
   vapor-compression heat pump, 45–46, 49, 52
   weak and strong absorbents, 111

J

Jet pump, 231, 237

K

Kalina Cycle, 55
Kirchoff's Rule, 124, 126, 128
Knots, 127

L

Latent heat
   enthalpy-concentration diagram, 29
   heat capacity ratio, 30
   heat pump effectiveness, 116
   refrigerator evaluation, 189
   two-stage absorption systems, 137
Lithium bromide/ammonia/water, solution fields, 34
Lithium bromide/water
   absorption heat pump evaluation, 94–98
   corrosion inhibition, 31
   double-effect systems, 167, 180
     compressor operation, 237
     COP, 199
   enthalpy-concentration diagram, 20
   heat pumping pressure levels, 32
   heat pump temperature lift, 193
   high-efficiency two-stage system, 158–159
   solution fields, 33–34

   two-stage system configuration, 148
   vapor pressure curves, 46, 89
Lorenz Cycle, 52
Low pressure systems, 255
Low-temperature flow rates, 236

M

Main components, 123
   graphic representation, 128
Mercury, 133
Mixtures, See Working fluid mixtures
Modes of operation, See Operating modes
Multistage absorption/compression/expansion
     systems, 231
   one compression/expansion machine, 233–235
   special configurations, 236–239
   three-stage configurations, 231–233
   two compression/expansion machines, 235
Multistage absorption systems, 123, 179
   COP, 187, 191, 194
     double-effect chiller, 195–196
     heat pump, 199–201
     heat pump transformer, 196–197, 202
     heat transformer, 201–202
     refrigeration, 198–199
     resorption cycle, 198
   decomposition and superposition rules, 194–195
   four-stage configurations, 186–187
   solid-vapor systems, 246–250
   three-stage units
     configurations, 181–186
     operating modes, 179–181

N

Non-zeotropic mixture, 22

O

Open systems, 255–256
Operating modes
   many temperature levels, 12–17
   three stage absorption units, 179–181
   three temperature levels, 6–12
   two-stage absorption/compression/expansion unit,
     208–213, 221–222
   two-stage absorption systems, 137–143
   two temperature levels, 2–6
Organic Rankine Cycles, 55
Overlapping temperatures, 170

P

Parasitic heat flow, 6
Part load performance, 283–286
Permutation operators, 174–177, 229
Planck's formula, 190
Potassium working fluid, 133
Power generation cycles
   cascade configurations, 133

power plant evaluation, 38, 42–44
steam power plant, 42–43
two-stage configuration, 231
Pressure
absorption/compression/expansion unit
adjustment, 219–220
thermochemical storage unit, 243
two-stage system evaluation
three levels, 151–154
two levels, 143–151
Pressure drop
assumptions, 124
graphic representation, 128
Pressure-enthalpy diagrams, 24–29
vapor-compression heat pump, 44
Pressure ratio, 49, 61
Pressure-temperature diagrams, 24
Pumping, solid adsorbents, 89, 241
Pump work
desorber-absorber heat exchange, 108
refrigerator evaluation, 190
single-stage heat transformer, 193
steam power plant, 42–43
two-stage absorption systems, 146–147, 152
vapor-compression heat pump with solution
circuit, 58–59, 214
water/LiBr absorption heat pump, 97
Purging device, 255

Q

Quadruple-effect configurations, 159–161
Quality factor
absorption heat pump, 115–118
compressor, 66, 68–69
refrigeration, 65

R

R12, 69
R22, 27, 69, 75
R22/R142b mixture, 24, 28, 48–50
R114, 71, 75
Rankine Cycle, 37–38
phase changes, 46–50
Rankine/Rankine Heat Pumps, 231
Rankine Sorption Cycle, 37, 50–52, 86–87, 190
Rectification, 98–103, 144–147
Reference system, 198
Refrigerants, See Working fluid mixtures; specific
refrigerants
Refrigeration, See also Absorption heat pumps;
Vapor-compression heat pumps
capacity, 116–117, 165–167, 225
cascade configurations, 133
COP, 5
multistage system, 198–199
Second Law-based determinations, 63–68
efficiency, 65, 67, 114
evaluation, 92, 188–191
power, 67–68

quality factor, 65
solid absorbents, 241, 246
Resorption cycle, 198

S

Safety considerations, 30, 32
Secondary components, 123
Second Law analysis, 1–2
COP determinations, 62–63
analytic equations, 70–75
exergy comparison, 63, 75–77
heat pumping, 68–69, 113–117
heat transformer, 118–121
refrigeration, 63–68
results for selected fluids, 69–70
four temperature heat pump transformer analysis,
17
sign combinations
many temperature levels, 12–17
three temperature levels, 6–12
two temperature levels, 2–6
single-stage heat transformer, 164–165
Set properties, 174–177, 228–229
Sign combinations
compression/absorption configurations, 210–213,
221
many temperature levels, 12–17
three temperature levels, 6–12
two-stage systems, 150–151, 210–213
two temperature levels, 2–6
Sign conventions, 1
Silicagel, 246, 255
Solar heat, 242
Solid absorbents, 89, 241, 244–246
Solid-vapor systems, 89, 241; See also Heat storage
compression/expansion machines, 251–253
COP, 250
equilibrium, 46
internal heat exchange, 247, 250, 251
multistage heat pumps, 246–250
quasi-continuous operation, 250–251
Solution circuits, 37
COP enhancement methods, 103 104
design rules, 127
double-effect chiller, 195
entropy production, 116
merging, 168
resorption cycle, 198
two-stage absorption systems, 140, 168–170
high-efficiency system, 159
vapor-compression heat pump, 38, 42, 52–61,
213–218
Solution fields, 24
single-stage heat pumps, 111
working fluid mixtures, 33–34
Solution heat exchanger, 98
Sorption storage devices, 241–243
Stage numbers, 126–127
Steam power plant, 38, 42–44, 255
Superheating

pressure-temperature diagram, 24
temperature-concentration diagram, 19
two-stage absorption/compression/expansion unit
    use, 225–227
Superposition, 126
    rules for, 194–195
Surface area allocation, 285

T

Temperature glide, 22, 46, 48, 283
    COP and, 48
    pressure-enthalpy diagram, 28
    two-stage absorption/compression/expansion
        system, 219
    vapor-compression heat pump, 49
        with solution circuit, 52, 55, 61
Temperature lift, 220, 283
    absorption/compression/expansion unit
        adjustment, 220
    cascade configurations, 133
    COP and, 201–202
    heat pump working fluid, 193
    heat transformer, 163–165
    internal heat exchange, 127, 155
    two-stage absorption systems, 140, 142, 161–163
Temperature-concentration diagrams, 19–24
Temperature-entropy diagram, 41
Thermochemical heat storage, See Heat storage
Thermodynamic diagrams, 19
    enthalpy-concentration, 19–21, 90–91, 145
    pressure-enthalpy, 24–29, 44
    pressure-temperature, 24
    temperature-concentration, 19–24
    temperature-entropy, 41
Thermodynamics, laws of, See First Law analysis;
        Second Law analysis
Three-pass heat exchanger, 98
Three-stage absorption systems
    configurations, 181–186
    operating modes, 179–181
Throttle
    turbine replacement, 37–38
    work values, 69–70, 73
Topological equivalence, 126, 131–132
Topping cycle, 133, 237
Toxicity, working fluids, 30, 89
Triple-effect unit, 159, 179, 199, 247
Trouton's rule, 31
Turbine, throttle replacement, 37–38
Two-stage absorption/compression/expansion
        systems
    classification, 207–208
        set properties, 228–229
        switching between configurations, 227–228
    COP, 218, 225
    fluid stream adjustments, 218–219
    internal heat exchange, 222–225
    multipurpose flexibility, 225
    operating modes, 208–213, 221–222
    pressure level adjustment, 219–220

simultaneous compression/absorption, 220
superheat use, 225–227
temperature lift adjustment, 220
Two-stage absorption systems, 135
    classification of, 135–137
        inter-configurational switching, 170–174
        set properties, 174–177
    COP, 147–148, 154
    efficiency, 137, 140
    enthalpy concentration diagram, 145
    evaluation
        three pressure levels, 151–154
        two pressure levels, 143–151
    heating/cooling capabilities, 165–167
    heat of rectification, 144–147
    heat pump transformer, 165
    heat transformers, 163–164
    high-efficiency systems, 158–161
    internal heat exchange, 137, 140, 154–157
    large temperature lifts, 161–163
    latent heat temperature dependence, 137
    operating modes, 137–143
    sign matrix, 150–151
    working fluid-solution stream connections,
        168–170

V

Vapor-compression heat pumps, 37, 83; See also
        Compression/expansion systems
    COP, 45, 49–50, 111, 217
        Second Law-based analysis, 62–77
    efficiency, 38
    exergy loss, 75–77
    heat and work sign combinations, 4–5, 11
    internal heat exchange, 45–46, 49, 52
    performance evaluation, 44–45, 111–112
        with internal combustion engine, 112
        R22/R142b system, 48–50
        with solution circuits, 55–61, 213–218
        steam power plant, 38, 42–44
    pressure-enthalpy diagram, 24, 44
    Rankine cycle, 37
    solution circuits, 38, 42, 52–61, 213–218
    special multistage configurations, 236–239
    two-stage configuration, 231
Vapor pressure curve
    absorbent/working fluid systems, 31–32, 46
    absorption heat pumps, 89
    pressure-temperature diagram, 24
    over zeolite, 244
Vapor pressure reduction, 55, 83
Vapor quality, 22, 37
Volumetric capacity, 30, 61

W

Water
    COP parameters, 72
    low-temperature flow rates, 236

Water chillers, 32, 90
Water vapor
  air-flow rates, 255
  rectification, 98–103
Window, 31
Work
  compressor, 44–45, 48, 216
  pump, See Pump work
  Rankine Cycles, 37
  Rankine Sorption Cycles, 50–52
  refrigeration COP, 65, 67
  sign conventions, 1; See also Sign
        combinations
  sorption storage system, 243
  throttling, 69–70, 73
Working fluid mixtures, 19
  absorption systems, 31–36
  azeotropes, 22–23, 36
  COP parameters, 69–77

double-effect cycle COP differences, 199
Organic Rankine Cycles, 55
properties of, 29–31
Rankine Cycle phase changes, 46–50
research areas, 36
thermodynamic diagrams, 19
  enthalpy-concentration, 19
  pressure-enthalpy, 24–29
  pressure-temperature, 24
  temperature-concentration, 19–24
toxicity, 30, 89
vapor-compression systems, 31
  with solution circuit, 52–61
vapor pressure curves, 55, 89

Z

Zeolite systems, 46, 255
Zeotropic mixture, 22, 46